CoE:
The Key to
Data-Driven
Manufacturing

CoE:
The Key to Data-Driven Manufacturing

Grant Vokey

Notice

The information presented in this publication is for the general education of the reader. Because neither the author nor the publisher has any control over the use of the information by the reader, both the author and the publisher disclaim any and all liability of any kind arising out of such use. The reader is expected to exercise sound and professional judgment in using any of the information presented in a particular application.

Additionally, neither the author nor the publisher has investigated or considered the effect of any patents on the ability of the reader to use any of the information in a particular application. The reader is responsible for reviewing any possible patents that may affect any particular use of the information presented.

Any references to commercial products in the work are cited as examples only. Neither the author nor the publisher endorses any referenced commercial product. Any trademarks or trade names referenced in this publication, even without specific indication thereof, belong to the respective owner of the mark or name and are protected by law. Neither the author nor the publisher makes any representation regarding the availability of any referenced commercial product at any time. The manufacturer's instructions on the use of any commercial product must be followed at all times, even if in conflict with the information in this publication.

The material and information contained in this book are for general information purposes only. Views and opinions expressed by the author(s) are solely their own and do not necessarily represent those of ISA.

Copyright © 2023 International Society of Automation (ISA)
All rights reserved.

Printed in the United States of America

ISBN-13: 978-1-64331-225-5 (print)
ISBN-13: 978-1-64331-226-2 (ePub)
ISBN-13: 978-1-64331-227-9 (Kindle)

No part of this work may be reproduced, stored in a retrieval system, or transmitted in any form or by any means, electronic, mechanical, photocopying, recording, or otherwise, without the prior written permission of the publisher.

ISA
3252 South Miami Blvd, Suite 102
Durham, North Carolina, USA 27703
www.isa.org

Library of Congress Cataloging-in-Publication Data
Names: Vokey, Grant.
Title: CoE : the key to data-driven manufacturing/Grant Vokey.
Description: Durham : International Society of Automation, 2023. | Includes bibliographical
    references and index.
Identifiers: LCCN 2023943092 (print) | ISBN: 9781643312255 (paperback)
LC record available at https://lccn.loc.gov/2023943092

Acknowledgments

I would like to thank the following people for their support and input to this book:

- Ankit Jain for his collaboration on sections of the book and his input into the pharmaceutical industry.

- Nityanand Singh for his expertise in all things IT and IT interfacing within the MOM-related systems.

- Liegh and the editors from the International Society of Automation, whose continued support has made me a better writer.

- Julie Fraser and Tech-Clarity for the use of the survey "The Manufacturing Data Challenge" (copyright 2020).

I would also like to thank Julie Fraser for her excellent feedback on the book and Tim Shaw and Dennis Brandl for their recommendations during the peer review.

Contents

About the Author

Grant Vokey is the principal consultant for Vokey Consulting. With 20 years of diverse manufacturing operations experience and an additional 15 years of integrating information technology (IT) systems into the manufacturing floor, he has developed a strong understanding of how manufacturing companies work and the information needed to operate at world-class levels.

Grant's experience, coupled with continuous training and 10 years as a Certified Operations Manager, has also provided him with an excellent understanding of industry best practices and best-in-class utilization of manufacturing execution systems (MES). Using this knowledge, he has been a subject matter expert for developing industry-leading MES applications/solutions, a program manager for multiple MES programs, and a lead consultant on implementations of MES in various verticals (electronics, industrial equipment, automotive manufacturing, and metal fabrication).

Grant has developed a reputation for providing sound, practical advice and direction that make a difference to his clients and the MES industry as a whole.

Preface

During a youth leadership course I took many years ago, I was introduced to a game that was intended to show the importance of good communication (many of you probably remember it). The game involves a group of people, and one person quietly tells the first person a message. Each person, in turn, whispers that message to the next person until the last person is given the message. The last person shares the message they received with the group, and the group compares that message to the original one. Like others in the group, I was amazed at how much of a difference there was between the first and last versions of the message.

Over the years, I have noticed a pattern of that kind of miscommunication in manufacturing companies. I have observed these issues throughout the companies, on both the business side and the manufacturing side. These issues have had a significant effect on initiatives for implementing any aspect of change management. However, from what I've seen, manufacturing operations management (MOM) seems to have more than its fair share of potential scenarios for miscommunication.

Why does this scenario play out so often in companies?

The specific answer to each independent observation is, of course, too narrow to be of value in a general context. But when I have stepped back from any particular issue, I have found that there were questions that could be meaningfully asked in a more general context.

In this book, I present some of the more general issues I found on the manufacturing floor and offer some ways to avoid or correct them. I primarily provide direction in creating a team within Manufacturing Operations whose purpose is minimizing the occurrences of these issues. Is it a lofty goal for a book? Maybe. Or maybe I might surprise you.

1

Introduction and Overview

Anyone who works in manufacturing quickly learns that change management is a big deal in the industry, to such an extent that consulting in change management in different industry segments provides a more than reasonable income for many. Think about how many consulting companies specialize in the Lean, Six Sigma, ISO-9000, or Theory of Constraints change management methodologies. Not to mention the number of people employed in manufacturing companies whose sole focus is change management as it relates to continuous improvement (CI). Also note that each of the mentioned change management methodologies started in manufacturing and was then recognized as being of value to the rest of the company. All of this indicates the importance of change management to manufacturing in general.

Yet many manufacturing company executives have expressed frustration with their company's inability to sustain CI for longer than the period of the original initiative to implement one of these change management methodologies. In addition, many senior operations managers have expressed some disappointment in the results of implementing any of the change management methodologies (or at least they expected considerably more out of the implementation than was achieved). In industry surveys as recent as 2020, up to 70% of companies have been disappointed with the results of their CI programs.

No, this is not yet another book on CI. I'm not going to tell you about a *fantastic new methodology* or that any particular methodology is better than another. Each methodology has a particular scope of operations it is meant to help improve, and each has its place.

I will say that it is not the particular methodology that is important (sorry to burst the bubbles of all those consultants); it is the combination of methodologies and the correct coordination of resources that make a difference. And to properly coordinate those resources, you must provide them with the "tools" in a program management context to enable them to develop. However, that raises a question regarding *how* to coordinate those resources.

My first book, *MES: An Operations Management Approach* (co-authored with Thomas Seubert), discussed using a manufacturing execution system (MES) to coordinate production and production support activities from an operations management perspective, and we provided some guidance on using the ISA-95 standards as a template for MES implementation. I'm happy that the book is now being reviewed for use by universities in master's-level courses for industrial engineering (specializing in manufacturing engineering). After spending more than 20 years implementing MES, I am surprised at how little growth there has been in understanding how to tie MES implementation to a company's CI program and in the ability of companies to manage these programs to be sustainable.

In this book, I provide an overview of the different characteristics of manufacturing management at the production floor level, how to use those characteristics for operations management, and how many of the latest trends in analysis and access to real-time data are actually of value. I also explain how to develop a coordinated approach to use these characteristics and some general "operations management tools" to create a specialized team to lead manufacturing companies into a long-term sustained program for continued improvement in manufacturing capability. This is not a step-by-step instruction because the detailed implementation of any sustained program is specific to each company, and this is not a book about quality management (there are many good books that cover that topic). I do, however, provide guidance on implementing a sustained program and explain some of the characteristics of manufacturing processes that enable the guidance that I provided.

Although the problems of change management are common to many industries, manufacturing industries have particular concerns. Not only do changes have to be made in the way the business operates (the business processes), but the same types of changes are needed at the production floor level as well (the manufacturing processes). Because of the detail to which manufacturing processes must be defined, the interactions between business processes and manufacturing processes, and the effect on the quality of even seemingly small changes, change management in manufacturing operations management (MOM) has a deeper level of complexity.

Many companies that have been repeatedly successful in implementing required changes have attributed that success to first implementing a formal process management program and a formal team to manage the *normalization* of processes in general, and then managing *both* the *changes in process* and the *process of change*.

In this book, I look at process management from a MOM perspective and cover the structure and implementation of a process management team within that MOM context.

In this chapter, I use the term MOM in a general context. In later chapters, I will get into more detail about the definition of MOM from an industry point of view.

Also, within that same general MOM context, many manufacturing companies have found that MES can play a significant role in the effectiveness of the process management team and of MOM in general. For that reason, this book also includes a general overview of MES functionality and how it supports MOM in process management as well as production efficiency and CI, and how an MES is a major component of successful companies' Industry 4.0 initiatives. As part of the Industry 4.0 discussion, I explain why creating an Industry 4.0 strategy is vital to manufacturing and how using data provided by smart sensors, machine learning, and automation is an integral part of an Industry 4.0 initiative.

Overview of Company Planning

To initiate planning, Operations senior management must understand their current state of affairs (capacity, process management capability, production planning, etc.). Hopefully, senior managers have been following the activity of the departments under them and already have that *current state* visibility. On a separate track of activities, Sales and Marketing will look at the company's current market position and work with Product Engineering to better understand the characteristics of the product line (current and upcoming), and they will determine what changes are needed to maintain or improve the market position. They will then develop an action plan for the company to achieve the desired changes in their market position. Once Sales and Marketing have their *plan of attack* for market gains and Operations understands the current state, it is time for them to come together for sales and operations planning (S&OP). They will determine what must change in the current state of operations over the course of the next year or more to support the sales and marketing plan, or they must determine what must change in the sales and marketing plan because of things Operations cannot support (e.g., as a result of capacity or resource issues). This is considerably simplified, but the result of S&OP is a plan to either hold a current market position or to make the gains desired and an action plan to update operations capability to support the sales and marketing plan.

After the plans are developed and approved, they are flowed down to department managers who simply execute the plan. Right? *(Um... ya... ok, sure)*. The first rule of company planning is that a plan is only good up to the point of starting to execute it. When the department managers start to act on the plan, a multitude of things can (and will) go awry.

After S&OP has developed the operations and the sales and marketing plans to move forward, senior sales and marketing managers will begin to execute their side of the plan, and senior operations managers will work with operational department managers to plan the steps for fulfilling the changes defined during S&OP. Again, there may be issues at the department level that must be reconciled to achieve the plans. That is why S&OP is actually an iterative process.

During the year, in addition to day-to-day activities, senior managers support department-level activities (either as part of a steering committee or by sponsoring the finances or both) to achieve the expected changes. As the year progresses, things change. Assumptions made during S&OP are sometimes found to be false, the company's priorities may change because of local or global events (e.g., the devastating economic effects of a global pandemic), or internal changes may drive different activities.

Although things can also go wrong from a sales and marketing plan perspective and a product engineering perspective, these issues are not under the control of manufacturing operations and, therefore, are beyond the scope of this book. From a manufacturing operations perspective, there are a few issues that can create problems.

One of the first issues is the accuracy of the original current state understanding of operations capacity. With multiple production lines (or multiple plants) and ongoing changes in operations staff, equipment, and products, there is a lot of potential variation in production activities that can cloud (or downright confuse) the current state of understanding. In addition, implementing the changes in operations may be more complicated than originally thought or planned, and required resources may not be available when needed. This is especially true if the changes require coordinating multiple departments. And, of course, there is the ongoing issue that these departments must also keep up with their normal day-to-day activities to get products out the door.

Overview of Operations Management for Manufacturing

The scope of operations management is very broad. It can include all aspects of a company's activities from receiving a customer's order through production, delivery, payment collection, after-market services and customer support (initial and long-term), warranty activities, and product servicing and upgrades while the customer is using the product. Within each of these overarching activities, a few functions become key to operations management in manufacturing. These functions include capacity management, inventory and resources management, control of execution of manufacturing, and assurance of the quality of the products manufactured or the services provided by the company. As resources (material, equipment, or people) are in limited supply (this is the basis of economics and business management), the ability to plan, allocate, and utilize resources efficiently differentiates successful companies from struggling companies. Companies must also execute change management with a high degree of efficiency. For an issue that requires change, Operations must determine the root cause, identify possible solutions, select the most effective solution, and quickly implement that solution, all of which takes resources (most of the time, the same resources used in day-to-day operations). The longer a problem exists, the more money a company is potentially losing. To anyone who has in-depth knowledge of operations management, this should not be new information.

A major concern that many operations managers have regarding executing the plans from S&OP and lower planning levels, as well as many of the daily decisions while executing these plans, is ensuring that they have the most up-to-date information available so that they are making well-informed decisions and accurately conveying the intended meaning of that information throughout the company. A key concept that is frequently overlooked is the relationship between the information used to make a decision and the accuracy and integrity of the data used to create that information. Some decisions can be made using information that is a few days old with little consequence. However, as processes become more complex, that relationship tends to become diluted, and (as introduced in the preface) the information being conveyed tends to drift from its intended message. This tendency becomes more significant as days-old data that one department uses to make decisions might be a couple of months old by the time the data is propagated throughout the company. In this regard, effectively collecting data and communicating the intended information in a timely manner can be paramount to a company's success.

In today's market, customers have become accustomed to highly customized products and the ability to change their orders at almost any time. Determining the details of which product to manufacture has become much more complex, and being able to deliver products in a short enough time frame has become much more difficult. From the perspective of a manufacturing company, the effort to deliver quality products—with the correct options, on time, and in the right quantity—requires knowledge of customer orders and quick updates to accommodate sudden changes in capacity. In day-to-day operations, unexpected changes in available resources are difficult to manage, and adding change management execution to the mix amplifies the importance of having knowledge of capacity and resources in real time. Whether managing day-to-day activities or implementing changes, when a manufacturing manager is trying to make the most effective use of resources (people, equipment, or material), unexpected changes (good or bad) are bad for planning. Either capacity is not available when needed and activities cannot be completed, or resources sit idle. Idle resources cost money to maintain, and there is a loss of opportunity when idle resources could have been used elsewhere. So, if capacity and resources are not planned accurately, it will cost the company in lost sales (customers go to another supplier) when demand cannot be met, or invested money is wasted as the estimated demand (reflected in the plans) fails to provide the expected return, thereby increasing the cost of production or reducing profits.

Figure 1-1 provides an overview of the scope of MOM and how information is created and made available throughout the different groups within MOM. Some of the reports

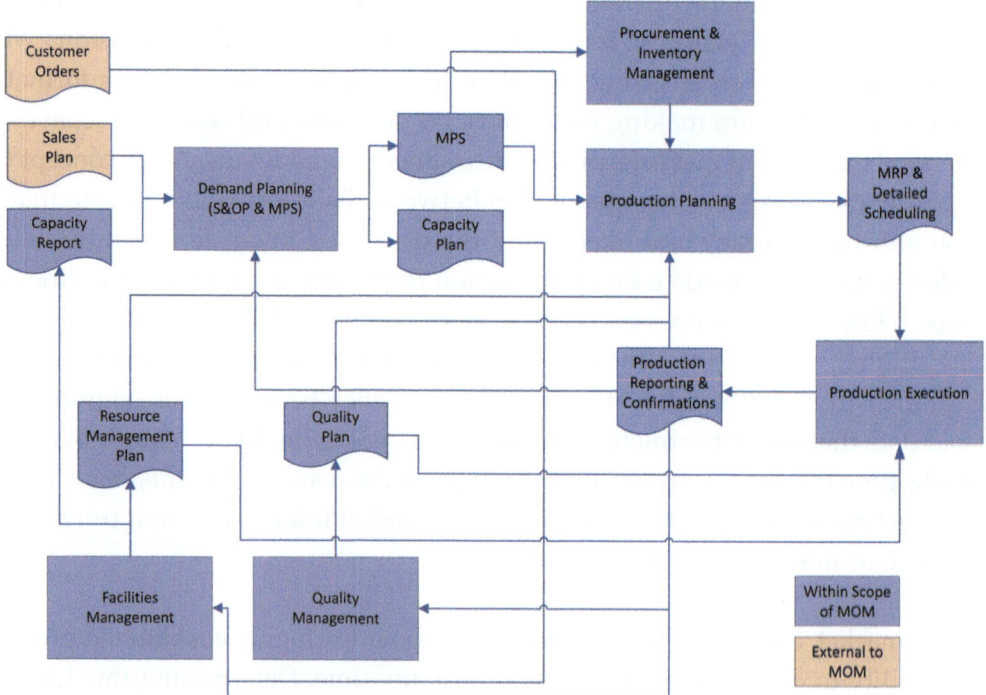

Figure 1-1. Scope of MOM and information transfer.

(or plans) that are used (e.g., the capacity report) do not have to be as current as others (the materials requirements planning—MRP, report and detailed schedule). Therefore, if the capacity report is a few days old (in some companies, it can be as much as a couple of months old), it may not have much of an impact on the capacity plan and the master production schedule (both are more long-range plans of change and scheduling) as an output of S&OP to facilities management or procurement. However, if the resource management plan (which likely includes personnel planning) used in production planning and at the MRP level is a few days old, equipment required for production may not be available (e.g., if it was inadvertently scheduled for maintenance or an upgrade), causing a major delay in production orders. Understanding how reports will be used, the data required for the reports, and the procedures for extracting the data and creating the reports are all important when looking at the timeliness requirement for the data and the cost associated with fulfilling that timeliness requirement (e.g., the cost of design for system interfaces).

Tools of the Trade in MOM

As I discuss in more detail in Chapter 2, "CoE and Data-Driven Management," there are a few *management tools* that operations managers use to stay on top of issues and manage the complexities of MOM. These management tools include reporting from manufacturing process management, regularly publishing reports, and reviewing these reports in regular, monthly operational review meetings. To prepare for these meetings, personnel can spend up to a week collecting data and summarizing the data into reports. Decisions are made during the meetings, and action plans are devised and acted upon until the next meeting. Issues that arise in these meetings include the reliability of the aged data that was collected, the consistency and integrity of the reporting, and the diversity of issues that require action to resolve. Although managing manufacturing in this manner in today's environment of fast data and immediate reports may seem archaic, many companies still operate this way.

The Need for Standardization

In addition to the concern about data (and its availability), there is also a need to minimize issues that require decision-making. A significant problem that operations managers face is being the sole person authorized (or able) to make decisions. This problem is further complicated when similar issues arise and must be re-evaluated, time and again, to ensure that the correct decision is made at that time. In some cases, this can result in "reinventing the wheel" as operations people reanalyze similar problems repeatedly to come up with more or less the same solution each time. As much as possible, the situations that require decision-making should be reduced either in the number of occurrences or in the complexity of the decisions to be made.

One way to solve this problem is standardization.

Standardization can take many forms, such as process standards, reporting standards, and standardized data collection. As a supplement to standardization, the production support team must develop the capability to record situational details whenever there are issues that the standard cannot support (e.g., the standard process must be changed for a specific issue); delays in recording those details can result in important information being lost.

Standardization Programs

Whenever a company engages in a significant standardization effort, it is important to have a team that is focused solely on the standardization activities. Although it might be easy to assign these activities to the manufacturing or quality engineering staff, adding standardization responsibilities on top of their other duties frequently results in delaying or dropping the standardization to take care of day-to-day operations issues. In addition, if a company has multiple manufacturing engineering teams (as in multiple plants), the implementation of *standards* will not be *standardized*. This can create the same problems of reinventing the wheel and complex decision-making that standardization was supposed to fix.

When a major initiative like implementing standardization is undertaken, the best practice is to use some form of program management. Although there are a few different program management methodologies, the important part of any program is to establish the standardized processes, reporting, and data and to integrate the overall management of these activities into the day-to-day operations management processes for long-term support. In this regard, it is best to develop processes for maintaining standardizations that are integrated into everyday operations and brought under the normal operations management structure while still keeping the dedicated standardization team to maintain continued progress in standardization and improvement. As part of integrating standardization into everyday operations, managing the program must include training the management and staff who form the standardization team as well as the managers who will use the reports and information created by the team. Many programs fail because, although the team recognizes the need for the staff to be trained to use the program, they fail to educate management on the program's value and the need to support it, as well as the need to incorporate standardization activities into their operations management.

As with any program, planning and implementing a standardization program inevitably requires access to resources that, at the same time, must continue to run their day-to-day manufacturing operations. As a result, companies typically hire a consultant to implement the program and get it running by developing the policies and procedures that define how to implement a standard process in general and developing a standardization team within the company to document, implement, and manage the standardization program. The team then ensures that the process collects the required data and publishes the information needed by operations management. Frequently overlooked during implementation is training operations management to use the reports and ensuring that there is a common understanding in interpreting the information the reports provide. After the policies and procedures are in place (and the knowledge becomes part of the normal operations management methodology), the consulting program manager is phased out, the standardization team becomes part of the operations management organization, and the operations manager becomes responsible for the growth of the program.

Although there are different titles for the team that manages standardization and process management, the term *Center of Excellence* (CoE) has taken hold in recent years. The function of the CoE can (and should) relate to all aspects of a company. However, this book focuses on the CoE function as it relates to MOM, and because an MES can be valuable to MOM, it also addresses the MOM functions that extend to implementing and using an MES.

Introduction to the Center of Excellence

Why should a company create a CoE, and what does it do?

A CoE has two primary functions:

1. The first is to provide the capability to fully model and document any of the business or manufacturing processes used within a company (in this case, any of the MOM processes).

    o Modeling includes the process capability and fallout, the data going into the process, and information needed by all stakeholders coming out of the process.

2. The second is to be a central source of knowledge for all other departments regarding what processes already exist, how they are measured for effectiveness, and which data is used to manage that effectiveness. It is also to help create and support a management plan to ensure continued improvement in efficiency and reduced production costs.

As the CoE matures in capability and knowledge, there are other roles it can provide. These roles are discussed in detail in other chapters.

With the information derived from operational reports, it is the role of department managers, with the help of the CoE, to track and deeply understand the factors that are impacting the effectiveness of their departments and then to determine a course of action based on the reporting results and factors that align with the company's capacity plan from S&OP. The department managers then assign the task priorities back to the CoE to work on to improve their departments' effectiveness.

In many companies, the CoE is initiated as part of the information technology (IT) organization. (This is particularly true in the MES industry.) However, the role of the CoE in this context is usually to ensure the application supports the process requirements as defined by the manufacturing engineer, but not the optimization of the process in which the application is used. Because the IT department has a limited understanding of manufacturing or business process management, problems can arise when the CoE is limited to IT knowledge (more on this in Chapter 2). To ensure that processes move toward optimization and are then repeatable to other departments, other production lines (in the context of manufacturing specifically), or other plants, the CoE must include members who have a holistic knowledge of operations management and, specifically, knowledge of process management methodology.

In Chapters 2, 3, and 4, I outline the roles and skills needed for a CoE, and I explain how the CoE is needed to support all levels of operations management up to the chief operating officer.

To understand the full context of the CoE to MOM, I include a more detailed comparison of *business* and *manufacturing* process management and why they should be treated differently.

The CoE and Industry 4.0: An Introduction

Part of the role of the CoE is to stay abreast of the changes in technologies that affect MOM. In Chapter 4, "Industry 4.0 and Data Mining," I describe using the CoE and MES as they pertain to Industry 4.0, data mining, and machine learning; I also explain why supporting these initiatives is important.

The main concept of Industry 4.0 is that a company's success depends highly on the effectiveness of processes used within the company. This effectiveness depends on proper process management, the quality of the information going into the processes, and the quality and timeliness of the data and information coming out of the processes.

Two factors that are not included in many Industry 4.0 initiatives is the knowledge of how to interpret the information coming out of the processes and the quality of the decisions that are made as a result of that interpretation. In this area of company management, the CoE will likely be the main source of knowledge and training regarding correctly interpreting that information. Many manufacturing managers are trained in the general process and quality management methodologies. The training they receive (either in school or on the job) frequently does not include training in fundamental data mining techniques or in interpreting the statistical or probability information that is becoming available with Industry 4.0. Developing a strong CoE can go a long way to improving the knowledge and capability of management and operational staff to be successful in the data-driven manufacturing world of Industry 4.0.

In Chapter 4, I also discuss the details of using the production-level data that can be available and bridging that data from the operational technology (OT) and controls engineering perspective to the information technology (IT) and relational database perspective. The chapter includes examples of condensing raw OT data to be stored at the IT level as well as an example of using basic data mining to understand the production events that are reflected in that data. In addition, I provide an overview of IT data that has been maintained within a relational database containing manufacturing operations data and an overview of interpreting that data into events on the production floor.

Chapter 5, "The CoE in Maintaining MOM/MES," includes a detailed explanation of MOM from the perspective of the ISA-95 standards and of how the standard can be used to help manufacturing operations in general. This chapter also includes a detailed look at the key issues a CoE should take into consideration when selecting, implementing, and maintaining an MES.

In Chapter 6, "MOM and the Functionality of an MES," I discuss in some detail the general functionality of an MES, compare the functionality of an MES to enterprise resource planning (ERP), and explain some of the differences between the MES functions that have been designed to support *process manufacturing* and those designed to support *discrete manufacturing*. I also discuss the characteristics of *batch manufacturing* and how it compares to process and discrete manufacturing from a data and an MES perspective.

Other topics addressed include the following:

- Using an MES as a tool to support production and MOM as a whole

- Points of concern for implementation strategies for an MES

- Considerations for selecting an MES application

- Managing an MES configuration

Finally, in this chapter, I discuss some examples of integration management from an MES to the enterprise level and down to the shop floor equipment. I also address using the ISA-95 standards to guide that integration and provide some input on best practices for integration.

The CoE and Strategic Planning

Chapter 7, "The CoE in Strategic Planning and Management," provides a high-level look at developing corporate strategy and a deeper discussion of fitting MOM into that strategy. As part of the deeper dive, I address the ways the CoE can (and should) play a significant role in developing a corporate strategy and in executing that strategy, at least at the MOM level.

As the CoE initiative matures and stabilizes within a company, the purpose and goals of the CoE will change as well. In Chapter 8, "Connecting the Dots," I look at how the CoE can be used as a source of knowledge and innovation. By understanding the importance of standardization and methodical management of improvement, the CoE can provide guidance and training to line-level manufacturing engineers on accessing the system information available to them via ERP, MES, or controls-level historians and on interpreting this information correctly. When this data is accessible, the CoE can then provide guidance on using the information to drive CI. In the chapter, I also illustrate that additional value can be driven by the CoE in a discussion on providing MOM-level analytics to senior management to be used for strategic planning.

It has been my observation over 40 years in manufacturing that many companies are not exploring all the opportunities available when using a properly developed CoE. Specifically in the manufacturing industry, a broad spectrum of skills is required, and efficiently coordinating these skills is critical to the successful operation of any manufacturing company and in executing initiatives such as Lean, CI, and business process management. Although coordinating skills and activities is important in any company, there is a complexity to this coordination that exists only at the manufacturing floor level within a manufacturing company.

It is my hope that people reading this book will gain an understanding of that complexity and some insight into how to effectively implement a CoE for the manufacturing floor, how to use the management and methodology tools that are available, and how to establish systems (people, processes, and technology) to successfully coordinate these skills and activities at the manufacturing floor level.

I hope you will find this book as informative to read as I have found it enjoyable to write.

2

CoE and Data-Driven Management

In Chapter 1, "Introduction and Overview," I reviewed the planning aspect of manufacturing operations management (MOM), reflected on the importance of having data that is up to date and accurately presents an understanding of the "current state of affairs" in manufacturing management at the manufacturing floor level and the senior management level, and started to introduce the basic functions of a CoE. In this chapter, I discuss the aspect of data integrity in more detail and go deeper into the function of a CoE from the perspectives of data and process integrity. I also examine the changes in operations management resulting from the increased availability of data.

When interpreting the data from a MOM perspective, it is important to also understand that the information presented in MOM reporting represents a physical context of operations. Work-in-process reports represent the status of physical products that are moving through the production floor, inventory reports represent the physical material that is on hand, and resource status reports represent human resources and physical equipment that are available on the floor with the equipment's actual physical state of operation. Part of ensuring that the reports provide accurate information regarding the physical status of manufacturing is to confirm that the data used to create these reports accurately models the physical entity it is representing. Of course, this is true for all aspects of manufacturing operations reporting, but reporting at the planning level (e.g., enterprise resource planning—ERP) is summarized and delayed to some extent (depending on the report) and therefore loses accuracy. Getting accurate data models for these entities is critical for the integrity of the information reported from planning and from order execution, quality, facilities, and inventory management.

When companies try to compare the operations of one plant to those of another (or even one line to another), it is also important to ensure that the data used in reporting is consistent between the lines or plants. Then there is a need to interpret both the data and the reports similarly. Relating back to Chapter 1, without implementing a standardization program, a comparison of reports from one plant to those of another would offer little credibility.

Just as important are the transition of data from one operations level system to another via information technology (IT) interfaces and the continued correct interpretation of that data from one system to another. The difficulty, as touched on earlier in this chapter, is this: As data moves from the shop floor level (the equipment and production execution levels) to the MOM IT and enterprise systems (MES, ERP, and advanced planning systems), it becomes more aggregated and summarized. This aggregation tends to change the context and the accuracy of the information created using that data. Each system has created data objects (the collection of data structures that are used to represent a particular entity) to define an entity (e.g., an operation name) in the context that is common to the level of user for a particular system.

Let's look at an example of a data object at different system levels.

In Figure 2-1, the term *Operation* on the ERP routing is compared to the term *Operation* in an MES. When defining Operation 1010 in the ERP, the data object would be defined to include all material, resources (and their utilization), and labor compared to the process *segment* of ASSEMBLY1 and INSPECTION1 of the MES routing. However, on

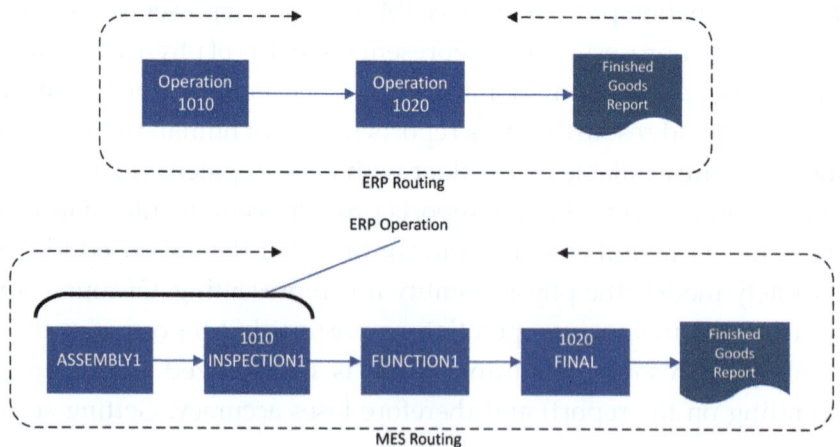

Figure 2-1. Operation data objects (ERP versus MES).

the MES routing, ASSEMBLY1 would be defined to include only the material, resource usage, and labor for that specific step in the routing, although the data objects would be referred to as an *Operation* in both systems. As data is transferred from one system to the other, the interface (and the people working with these systems) would have to recognize and properly interpret these differences.

The tendency for data (and the information created) to *drift* in integrity becomes greater if the company has multiple plants and the context of the data changes slightly from plant to plant.

As a result, it is important that the team designing the data models and interfaces for the system also have an in-depth understanding of the processes the systems are supporting. In addition, the team that is designing the processes that support the business and product manufacturing should have an in-depth understanding of the general information that will be needed to support the processes (from an operations point of view) and the data models that will be needed to create that information. To prevent confusion within the many groups of operations, a best practice within systems management is to develop a "data dictionary" to maintain the correct interpretations within the company.

In a recent industrial survey conducted by Tech-Clarity titled "The Manufacturing Data Challenge" (copyright 2020), of the 300 companies surveyed, 80% listed "consistent management of data in all plants" and "integration between equipment, plant, and enterprise" as being "important" or "critical" to data management. Of course, they are also critical to process management and business management. Ensuring that data from manufacturing and the reports created from that data are consistent throughout the manufacturing floor and across all plants requires implementing standards as discussed in Chapter 1. These include standards in the data models that represent the manufacturing floors, standards in the information or reports that are created to inform operations personnel of the "current state of affairs," and standards in the management training to interpret those reports. Standardizing the data consistently across multiple plants and keeping it that way requires a dedicated team (i.e., a CoE).

What Is a Center of Excellence?

In Chapter 1, I introduced the CoE from the context of manufacturing operations. It is important to understand that a CoE for one industry (e.g., manufacturing) will be different from a CoE for another industry (e.g., consulting in general). In consulting, a CoE might be the expert-level knowledge for the field in which the consultants are consulting (e.g., experts at the inner workings of a particular MES application) and

The mistake many companies make is managing the implementation of these initiatives as an independent project.

might be highly IT centric. In contrast, a CoE for a manufacturing company requires expert-level knowledge within the company's particular sector of manufacturing (e.g., electronics manufacturing) and may be highly process management centric.

The concept of a CoE is not really new, particularly to manufacturing companies. When a company takes on a new initiative (e.g., Lean or ISO-9000), it typically creates a team to act as the focal point of the initiative. The team members are the ones who are trained in the concepts of the new initiative and who define (with management's help) what that initiative is going to look like within the company. They become the internal experts on that initiative and drive the change management process defined by the initiative within the company. However, *the mistake many companies make is managing the implementation of these initiatives as an independent project*, and when the implementation is complete (e.g., the company gains ISO-9000 certification), the project team is frequently dismantled and the people who gained the expert-level knowledge go back to their day jobs. Then when the next initiative is undertaken, the company creates another team (that may or may not include members of the previous team) to drive the new initiative.

A skill set is needed to lead a company through change of any sort, and *that skill set will be needed for each and every initiative*.

The issue is that not only must these people become experts in the subject matter for the initiative but they also must become experts (through experience) in change management in general. A skill set is needed to lead a company through change of any sort, and *that skill set will be needed for each and every initiative* that the company takes on. In addition, for the most part, there is considerable similarity in the process of implementing these initiatives. Although there is a difference between implementing Lean and implementing ISO-9000, in each initiative, there is a need to map out the current understanding of a company's state, define the new vision of the initiative, and define the steps required to move from "the current understanding" to the "new vision." And, if the company is implementing properly (as part of a program and not as an individual project), it also must move the initiative from the "implementation" phase to the "business-as-usual" phase. There is also a difference between managing changes within a process using Lean (something a manufacturing engineer would handle as a process owner to implement *changes within* a company's management system) and managing changing the fundamental way a company operates (changing a company to a Lean mindset and implementing fundamental *changes to* the company's management system). The first example is a function of "business-as-usual" and does not need the skills of a CoE, whereas the second example is a fundamental role of the CoE.

CoE Development

Depending on a company's requirements, the CoE team will have a few roles for its members to handle. In general, the CoE will be responsible for documenting and managing the processes that are common to larger areas (multiple lines, production of multiple products, and across multiple plants) as covered in Chapter 1. The CoE will also provide the knowledge of the process and data model management as well as the knowledge of the method of *defining* the processes and the policies for *managing* these processes. In addition, the CoE will have ownership of what are referred to as the *look-and-feel policies* that determine how documentation will be presented and the terminology that will be used within the documentation.

Skills and Structure of a CoE

So, what exactly is a CoE, and how does it relate to MOM? A CoE is simply a team of people who have exceptional skills in a particular area of concern and are responsible for (1) defining the best practices within the context of the area of concern and (2) are responsible for ensuring continued improvement in the area of concern given improvements in knowledge and understanding, and improvements in technology that may affect the area of concern. For example, a consulting firm's CoE would develop expertise in configuring and using a system like an MES as it pertains to a broad range of manufacturing industries. The CoE team for the consulting firm may not have expertise in a specific industry, but given detailed input from industry experts, they can define a configuration that will work for any company in that industry. However, it may not be an optimal configuration for any specific company. In this scenario, the consulting CoE relies on the quality of the information provided to them by the industry (or company) experts. In an implementation project, the consulting firm relies on the manufacturing and quality engineers and assumes that the engineers understand the context of the MES.

The context of the CoE changes in a manufacturing company. The CoE is responsible for establishing the company's best practices in the area of managing a manufacturing environment. They are not necessarily establishing the manufacturing processes (that is the responsibility of the manufacturing and quality engineers), but establishing the method and scope of defining the manufacturing processes and the activities and methods to improve them. In addition, the company's CoE establishes the company's best practices for using available technology like an MES (both the MES within the company and any MES that are new on the market) to support these processes within the company. The CoE team may not have the broad understanding of a consulting firm, but they are very knowledgeable about how the company is using a system such as an MES. In this scenario, the internal team is reliant on the quality of the information provided about the system.

When implementing a CoE, it is important to understand that the skills required in the CoE must be based on the scope of work and to recognize that as the group's skills mature, its scope—and therefore its skills and structure—will change as its role in the company changes.

Frequently, the implementation of a CoE (at the MOM level) is started in a single plant by a manager who recognizes the value of standardization and has the leadership to guide staff through the initial steps, or it is started by the IT group, which recognizes that supporting a wide variety of processes with a single system is difficult (my experience is usually the latter,) and the IT group spends a lot of time pushing for any aspect of standardization (not necessarily the best aspect of standardization) if only to minimize the demand for supporting the system. Unfortunately, an IT-driven CoE trying to institute standardized processes has the effect of "the tail trying to wag the dog" and is met with much resistance. And although it is possible to initiate the CoE implementation from a single plant, the direction can be limited, and if or when the scope of the team extends to other plants, there is a strong possibility that all the processes that have been standardized within the limitations of the single plant must be revisited. As a result, it is best practice to initiate a CoE implementation as a centralized team. However, regardless of best practice, depending on the circumstances, it may be better to establish some aspect of process standardization within a single plant as long as there is recognition that any newly developed "standard processes" must be updated when the scope of the CoE changes which is the general premise of continuous improvement (CI).

Real-Life Experience

Before the company I was working for had considered the need for a CoE, our central team that was supporting the MES rollout globally (44 plants in North America and Europe) was responsible for developing configuration and "add-on" customization for each of the plants. Despite our frequent requests to normalize the customizations, each plant had made it clear that its processes needed individual support. However, as the plant-level manufacturing teams worked through their CI cycles, we found that, over time, some plants were requesting functionality that was already developed and used to support other plants a few months prior. It may have taken a little longer, but it was determined that the individual plants were independently converging on similar practices, and if the concept of a centralized knowledge base had been accepted at the plant level, the time needed to converge on a single "company best practice" could have been drastically reduced. Furthermore, this could have saved the company thousands of dollars in custom development that was only used for a short period and could have saved the plants more money in analysis time to come to the same conclusion.

Determining the CoE Structure

In the CoE initiation phase, there must be a lot of flexibility. Primarily, the CoE will have management and common skills available centrally within the company and will develop a knowledge base that is stored centrally. Often companies starting a CoE initiative maintain the knowledge base local to each plant or sometimes even to each line (some manufacturing engineers consider this information to be a competitive advantage for their personal career path), resulting in pockets of knowledge. On the other hand, plants may develop a knowledge base that is local to each plant without knowing other knowledge bases exist. As discussed in Chapter 1, this leads to a lack of standardization across plants.

Real-Life Experience

In the first initiative of developing a CoE (also my introduction to being a member of a central CoE), the company agreed to have two or three members from each plant in North America travel to the head office for an initial one-week meeting of what was to become the greater CoE team. A total of 30 people attended the meeting, and although initiated via the IT team, we had the good sense to recognize that the scope had to include the manufacturing and quality engineers as well. On the first day, each plant was required to present its processes for supporting an assemble-to-order (ATO) manufacturing model, and the enterprise applications (ERP), product lifecycle management (PLM), service-oriented application (SOA), and MES (I was there for SOA and MES, which was central to supporting the processes) were each to present their capabilities to support ATO. The presentations were eye-opening and proved that the meeting was required. Each plant defined a different "global standard process," and each enterprise application team presented what it called its "global standard configuration," which was based on a particular plant's processes. However, three different plants were used to model each application's "global standard," and the processes for each plant were different from each other in terminology, reporting, and some basic process support.

During the remainder of the week, we worked through defining a single process that all plants would implement, and we recorded the requirements for the applications to support the process. The requirements made no assumptions regarding which application would support specific requirements, and the IT teams agreed that in designing the enterprise-level scope of support, each application would take on supporting a requirement regardless of what design work was already in place.

As the configuration and interfaces were being developed for the applications, each manufacturing engineering team went back to its plant to start making changes to its lines to support the new processes, including training its production operators on the process and terminology. After the rollout started, we ran into only one problem with the configuration of the enterprise system as a whole, and the remaining plants rolled out without incident (one of the smoothest rollouts ever for the company).

The main factor of success in this project was having the input of all stakeholders (IT and the plant engineering teams) in the initial design of the fundamental processes and the requirements for the processes. An additional factor was the openness to starting from scratch in all design and development aspects and developing functionality and interfaces to provide *optimal* support for both the processes and the integrity of the data.

From the CoE perspective, the method the team used to create the fundamental processes became the model for other initiatives, and the agreements that the IT teams made in the initial design to support requirements became the policies that guided the design teams for several years.

An additional concern when structuring the CoE is the need to ensure that all stakeholders take ownership of the knowledge base. Unless there is a direct link between the central knowledge base and the plant-level manufacturing/quality engineering teams, there is still a high propensity for the plant-level teams to reject the information in the knowledge base (or ignore it), resulting from a "not invented here" mentality and continuing with the concept that "they are different."

Therefore, while developing the CoE structure, it is critical that the plant-level manufacturing and quality engineering teams are directly engaged in developing the knowledge base and, therefore, have at least part ownership of it. As shown in Figure 2-2, it makes sense to establish a common central team that maintains the leadership (or program management) role that establishes and maintains the direction of the CoE in general and includes staffing with knowledge of the required IT infrastructure and senior-level process engineering. The central team should also include an MES specialist and a data science specialist (these may be added later). There should also be representation from the manufacturing, quality and controls engineering disciplines for each of the plants (or lines within the plant, depending on the size of the

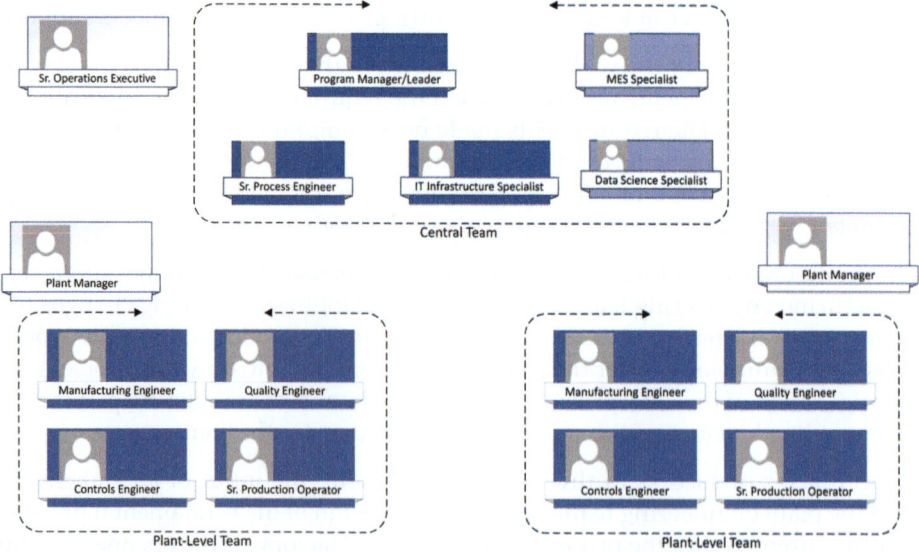

Figure 2-2. Example of initial CoE structure (central and plant-level teams).

manufacturing environment). As the role of the CoE in general changes within the company, it will likely have a greater influence on the workings of manufacturing operations. As a result, it is recommended that the central team report to the senior-most position within the company's operations. Because the plant's local teams will influence the entire plant's operations, each of

> The key to a well-developed CoE is not only for the team management to work with the company's senior management and plan for the roles senior management wants the CoE to fulfill, but also for the program manager to advocate for expanding the team's scope.

the local teams should report to the plant manager (or senior-most position within the plant operations). As a result of the split in responsibility for managing the CoE in general, it is essential to provide training/education on the CoE benefits to the operations management at the head office and the plant-level management. The plant-level engineering staff will be the main users of the centralized knowledge base, so the plant-level managers will need to be significant sponsors. Also, as a result of their influence at plant-level operations, it is important to ensure that plant-level teams participate in developing the CoE knowledgebase. At no time should the plant-level teams perceive the central team as "dictating" CoE activities.

One issue that may be encountered when implementing a CoE is that it takes time to create the skill sets, standards, methodology, and management of the CoE as an individual initiative. Fortunately, there are other initiatives that can be introduced and implemented as part of the CoE program. The ISO-9000 initiative is of value because it provides a management structure to build out the responsibilities of a CoE. Many of the steps needed to achieve ISO-9000 certification are the same steps a CoE must perform for the function of process mapping and standardization. The additional benefit of the ISO-9000 standard is that it establishes the requirement to audit the performance of and adherence to defined processes and to actively manage improving a company's capability. This provides guidelines for the CoE program and the ISO-9000 program that benefit both. In addition, the Lean initiative guides the actual process of mapping steps and therefore gives the CoE instructions on how to fully map a process. Therefore, it makes sense to develop the capability of a CoE and use the CoE initiative to drive and implement either ISO-9000 or Lean.

The CoE Skills: Early Stages

As mentioned in Chapter 1, the skills needed within a CoE depend on the scope of its responsibility and on the level of group maturity in process management. In the CoE, each defined skill is not attributed to a single person. In many cases, one person will have multiple defined skills. However, the team must be careful to avoid placing a considerable workload on that person. Depending on the manufacturing environment, several skills will be required initially. Each skill is outlined in the following list, with an explanation of why it is required.

- **Process mapping** – This is needed to be able to document the current "as-is" processes on both the business and manufacturing sides in detail. An understanding of process mapping is necessary not only to understand the main process but also to include any deviations that occur and to document the reasons for these deviations (if known). The skill set for process mapping is similar to that for value engineering. This is also a skill an experienced Green Belt in Lean would have.

- **Manufacturing engineering** – As a contributor to process mapping, the process map must include the details of equipment operations, as well as the interactions between different pieces of equipment and between the operators and the equipment used to manufacture the product. Manufacturing engineering plays a critical role in providing these details. They include the purpose of each piece of equipment, the acceptance criteria for the product's characteristics after the equipment is used, the settings for the equipment, and the data that is available from the equipment. Details must be provided for both manual and automated equipment with any specifics for production operator interaction.

- **Quality engineering** – The quality engineering team has the best understanding of a product's performance characteristics, the issues that are being defined for inspection, and the product test parameters that must be verified. This information is key in defining which data must be collected for the products' quality assurance and in understanding the reporting needed to verify that a product is ready to ship.

- **Controls programming** – For companies that use automation (which includes most companies these days), the equipment has usually been defined to use the data from sensors internally to ensure that the automated process is functioning properly. Much of this data is not programmed to be made available to external systems. As processes are mapped, it is also important to record the additional data that each piece of equipment has available internally. As process management matures, changes in controls programming to expose this data externally would be an advantage.

- **Data mining/analysis methodology** – Although this skill is not needed immediately, understanding the processes used for manufacturing and interpreting the data back to the processes is a unique skill set. Being able to collect data and create meaningful interpretations to properly understand production events is very important.

- **Document management methodology** – Many companies use a document management system to maintain process mappings. However, many of these systems are not used properly and are not organized in a manner that is useful to manufacturing staff and management. An additional gap in using these systems is the function of revision control. Knowledge of documentation management methodology requires not only maintaining the correct organization and revision but also developing the look/feel policies and a company-wide glossary of terminology.

- **IT infrastructure specialist** – The key requirement of the documentation process is to be able to define how a process is to be measured and how to report on the performance of a process. As a result, there will be a need to record data during the process execution and publish the information derived from that data. Even if process data is recorded in a spreadsheet, the knowledge to do that will be required.

- **Production planning** – When looking at the processes that support manufacturing, the team will also have to look at the processes that create and support production orders and support the movement of inventory onto the production floor (sometimes referred to as *line-side stocking* or LSS). If the data feeding these support processes is not correct, the initiation of any production run will be flawed from the start. Ensuring that correct data is being used for material requirements planning, just-in-time, and detailed scheduling is important to ensure that production continues as planned.

- **Program manager/leader** – As with any initiative, one person must take ownership of the initiative and act as the planner, manager/leader, and facilitator. It is vital to have a program manager who understands not only the phases of a process mapping initiative but also MOM in general.

As I mentioned, these skills will be needed at the start of the CoE initiative. However, it is also important to understand and plan for how the CoE's role within the company will change and what skill sets will change (or be added) as the initiative matures. Depending on the primary drivers of capability within the CoE, the skill set could include Lean and/or Six Sigma Black Belts or ISO 9000 internal auditors. The key to a well-developed CoE is not only for the team management to work with the company's senior management and plan the roles senior management wants the CoE to fulfill but also for the program manager to advocate for expanding the team's scope. Although this may sound obvious, in my experience, many companies that implement a CoE fail to plan for growth. I cover the roles a CoE can take in more detail in later chapters.

CoE Activities

As stated earlier, a CoE must undertake a lot of activity in the initiation phase of the program. In this section, I dive deeper into these activities and explain them in a manufacturing operations context. Most of the initial activities of a CoE come down to three main functions: documenting processes, normalizing these processes, and ensuring the repeatable measurement of these processes.

Documenting the Processes

The depth of process documentation will depend on whether the processes are business-level processes or manufacturing processes, and there is a difference.

When viewing processes from a CoE perspective, a few key concerns must be addressed to properly manage the processes in general, including the following:

- Defining and documenting the full scope of the process.

- Defining the criteria for successfully completing the process.

  o This includes both transactional and reporting outputs.

- Defining the ability to measure if a process is in control or not.

- Defining who has ownership of the process. Who is accountable for completion, and who is responsible for ensuring compliance?

Business-level processes are transactional (e.g., releasing a production order or an inventory move) or documentational in nature (e.g., recording the details of inventory receipt, recording details of an order transaction, creating a report for equipment calibration). These business-level processes are also relatively simple to follow, and as long as the data is correct and steps are followed, the output (the completion of a transaction or the availability of a report) is relatively easy to verify.

Manufacturing-level processes, on the other hand, have all the same points listed in the business-level processes, but they are also highly procedural (specific steps for assembling a product); are very detailed (may identify activity for each component on a product); have a strong relationship with physical activity and are, of course, related to the physical output of a product; and frequently include reliance on interactions with external sources of activity (as in working with tools/equipment). All these points can have an impact on the quality of the output (i.e., the quality of the product).

In manufacturing, the availability of data and the ability to analyze and report on that data in real time have become critical in managing day-to-day operations and initiatives for CI. In developing the capability to collect, analyze, and report on operations, the MES has become one of the most central

> The availability of data and the ability to analyze and report on that data in real time has become critical in managing day-to-day operations and initiatives for CI.

systems in supporting MOM. With this in mind, the remainder of this book will reflect on using an MES as the primary system for supporting a CoE and manufacturing operations.

The ISA Four Pillars of MOM Activity Model

To help with understanding the scope of systems developed to support MOM, the MES/MOM industry (guided by ISA) has categorized the MOM functions into four areas called *the four pillars of MOM* (Figure 2-3).

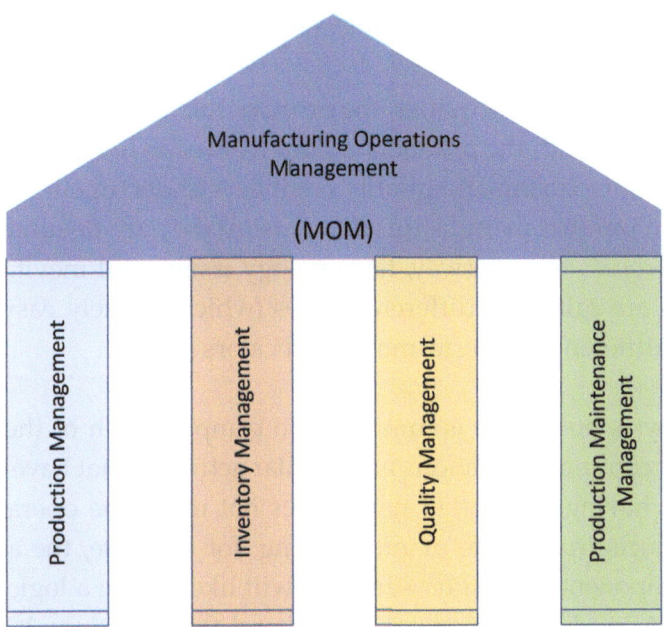

Figure 2-3. The four pillars of MOM.

The four pillars are the main functions of management for the production floor. They consist of the following:

1. **Production management** – Covering all activities to define, measure, and analyze the production planning and execution

2. **Quality management** – Covering all activities for executing quality plans, including testing, inspecting, and performing capability analysis

3. **Inventory management** – Covering the activities of ensuring the availability of correct material feeding the production floor and the correct availability of finished goods coming from the production floor, and any intermediary inventory management for materials and work in process (WIP) (e.g., managing moisture-sensitive or time-sensitive inventory)

4. **Production maintenance management** – Ensuring the ongoing availability of production equipment and tools

The four pillars will be covered in greater detail in Chapter 5.

Although initially the scope of a CoE will be investigating mostly from a production point of view, it is important to keep in mind that the CoE will likely grow to include the full scope of MOM represented by sections of the four pillars. Each of these sections will have its own set of processes that affect the manufacturing performance. Any one of these pillars can have a significant impact on MOM effectiveness.

In the initial stages of a CoE's activities, the primary role will be to document the major processes that interact with the production floor from each line (and each plant) and to work with the manufacturing and quality engineers to document the manufacturing processes for each product manufactured (or ensure they are documented). While the documentation process is underway, terminology issues will inevitably arise, where similar activities are called by different names (which is likely easy to resolve) and measured using different key performance indicators (KPIs).

During this analysis process, it is important to compare each of the individual processes to identify common operations and similar activities that have different names, recognizing that having a different name does not mean the operation is different. With many products in discrete manufacturing, for example, the assembly process of individual components and subassemblies will likely have a logical breakdown of subprocesses, tests, and inspection operations and therefore may have the ability of using a common naming structure.

It is the CoE's responsibility to work through these differences to understand the reason for the varying names and KPIs. In some cases, the differences may be a result of different views of the same measurement and resolving the issue will be easy (just choose one name and go with it). In other cases, differences arise from plants (or production lines) trying to achieve different goals. If this is the case, resolution may not be as easy, or there may simply be a need for multiple KPIs. The problem with multiple KPIs is that they can frequently conflict with each other. The best practice in managing processes is to reduce the number of KPIs that are measured at any time and to have a specific reason (trying to achieve a particular goal) for each of them. When the goal is achieved (e.g., as a result of a CI initiative), then the KPI can be changed as long as the gains from the original KPI are not lost. Although it is not required, this initial stage of CoE development can be included with an initiative for the quality management program ISO-9000 certification or a Lean implementation, where these activities are also required. Because these initiatives are similar, some companies will create a pseudo-CoE team as part of an ISO-9000 or Lean initiative. The company should then maintain the pseudo-CoE team after these initiatives are completed and formalize its responsibility to the full scope of a CoE.

Normalizing the Processes

After the processes have been documented, the next step is to work with the production and planning staff to ensure that the documented processes are followed consistently across all lines and plants, and that the processes are repeatable. A benefit of tying the CoE initiative with something like ISO-9000 is that the internal process audits performed as part of the ISO-9000 program can also be used to verify the consistency of process execution needed for normalization.

One aspect of MOM that is widely recognized in both process and quality management is that variation (deviation from a defined process or product characteristic) is bad. As discussed in Chapter 1, variation can cause concern with resource utilization when staff analyzes similar issues repeatedly to determine the cause of problems. Another reason variation is bad is that it hides the root cause of an issue by contributing multiple potential sources that may or may not be the root cause. By ensuring repeatability and consistency in following a process, production and planning staff help to keep variation to a minimum, and having fewer issues to eliminate as a root cause makes the analysis easier. Even if there is no actual problem other than the need to improve throughput or efficiency, it is still important to maintain consistency as it makes it easier to determine where the constraint of the process is and understand what to change to improve the constraint. So, during the initial phase of a CoE program, it is not critical to establish "an optimized" process (that will come later). As stated, the primary purpose is consistency in performance. This subject is covered in more depth in Chapter 8.

Ability to Measure the Processes

In determining how to measure the performance of a process, the CoE will also determine the data requirements that must be recorded in an effort to calculate the KPIs that are used as the primary measurements of success. When the processes have been documented (or, better still, *while* the processes are being documented), the CoE should also be looking at a means to record and store data. Admittedly, for manufacturing processes, I'm partial to using an MES for this purpose (this is what the application was designed for). Storing the required data in a historian (e.g., if the process is highly automated) or a Microsoft Excel spreadsheet may suffice to start with. The key point in measuring a process is having the data recorded, maintained, and frequently monitored and reported on to make the effort of measuring valuable.

There are two primary modes of failure in managing process measurement. The first mode of failure is not defining the data to be recorded and not creating a method for recording data for use in process management. Keep in mind that if you have no data to record, you have nothing to measure the process performance with, and you cannot improve what you do not measure. The second mode of failure is recording data in some manner but not incorporating that data into a regular report on the process performance that the data was originally designed to measure (i.e., recording data "just in case" but not reporting or using it as input for analysis). Many times, manufacturing and quality engineers have spent weeks monitoring a process and collecting data manually that has already been recorded via a historian or test system. At this time in CoE development, it is also important to start training manufacturing management on the importance of process normalization and calling on management to ask for updates on process performance as part of the monthly operational review meetings.

An additional scope of process management that is frequently overlooked is to monitor and measure the capabilities of MOM for general process management. Some companies do a good job of measuring their actual processes but poorly measure the "process of" process management. For this purpose, the more advanced companies are recognizing the value of the process capability maturity model (PCMM) as an established method of measuring process management in general. Capability maturity modeling (CMM) was originally established in the software development industry for managing the improvement in software development lifecycle management. It has been recognized by business process management (BPM) organizations as an effective tool for managing BPM generally. The factors that highlight good process management in business modeling are just as effective in managing manufacturing process improvement as well. More on this in Chapter 3, "Process Management, Continuous Improvement, and the CoE."

Building Data-Driven Management

Over the last several decades, there have been studies and reviews on the decision-making process of operations management. A consistent obstruction to decision-making within operations management has been a lack of data (or, more precisely, lack of information). In many circumstances, the data needed was limited, it contained too many interacting variables and was difficult to decipher, and it took too long to gather and analyze to be of value when managers were faced with making decisions quickly. As a result, many managers were put in the position of basing decisions more on intuition than on hard facts (i.e., supporting information). As discussed in Chapter 1, there are techniques for reducing uncertainty in decision-making, such as standardization and developing a company's set of best practices. These best practices help to provide mental models (a general understanding of what should be happening that a manager can think about as a reference) of production activity and guide the quick analysis that managers engage in to help with intuitive decision-making. Standardizations help to reduce variability, which further helps to reduce the complexity of operational decision-making. Operations managers have been taught for years to use these mental models and standardized processes as management tools to help with the decision-making process. In manufacturing operations in particular, managers are faced with decisions multiple times a day on issues ranging from order planning to shipping schedules and from the containment of product quality issues to changes in resource availability that result from equipment breakdown or employee absence. In light of all the decision-making that is required in MOM, it is no wonder that managers rely on standardization and best practice models to get through the day.

It is important, however, to remember that these "management tools" are not designed *for analysis to derive information*; rather, they are a technique *to deal with the lack of information*.

Now enter Smart Manufacturing, Industry 4.0, the Industrial Internet of Things (IIoT), IT/OT (information technology/operational technology) convergence, and many other technology buzzwords. The premise of this entire change in industry is that with IIoT and developments in analytics from machine learning techniques, the "lack of information" is no longer a decision-making problem. In fact, with the use of IIoT and real-time analytics, there is now the possibility of having too much information. The risk in this new situation is managers having a hard time deciphering the mass of information that can be derived from the data available and then determining what is *really relevant* at that time. The issue now is giving managers the tools to draw *relevant* information and to get them to supplement the previous tools of intuition and best practice modeling with the interpretation of this new information. It is also important to understand how the new information can fit into MOM activities. This introduces the next discussion in the value of a CoE.

With the combination of skills included in the CoE, a considerable amount of knowledge builds in terms of how the production floor is represented within the data models

Without a clear understanding of the processes, developing a valid budget is a little more than an educated guess and holding to that budget has a high dependency on luck.

and in understanding how the data is used in generating the reports. With that knowledge, the CoE then becomes very experienced in interpreting the reports back to the events and proceedings of manufacturing operations.

It now becomes important for management to work with members of the CoE to gain a similar level of interpretation skills. With practice, management then becomes much more confident that the reporting is accurate and up to date. Managers should understand where the data is coming from and how the data is being presented in reports. Only when managers have this level of understanding will they have confidence in what is being presented.

Using Data for Manufacturing Management

Two significant financial responsibilities that managers have are related to budgeting. One of these responsibilities is estimating what it is going to cost to run their departments (building the budget). The second responsibility is to ensure that the department stays (reasonably) within that estimated budget. The issue with the budgeting process is that it is primarily based on a rough scope of the activities within the department for day-to-day operations, but the budget also should include a rough estimate of the additional cost of extra initiatives and training that are expected as a result of strategic planning (more on this in Chapter 7). For manufacturing managers in particular, the budget includes the additional scope of understanding the direct and the overhead labor requirements for manufacturing operations and the "cost of manufacturing" for each of the products being produced (depending on total resource utilization per production unit). Without a clear understanding of the processes used during direct manufacturing activities and the processes supporting manufacturing within their overhead, developing a valid budget is little more than an educated guess, and holding to that budget is highly dependent on luck.

Part of the outcome of the CoE's efforts in process modeling is that the operations manager can now gain a much clearer understanding of the resources that are needed not only for developing a better picture of the cost of manufacturing but also for the overhead activities of special initiatives that are needed to improve budget estimation and expenditure control.

As originally discussed in Chapter 1, having well-documented processes that are followed significantly reduces the variation that causes variance in quality and in budgeted expenditure. In the next couple of sections, I look deeper into other areas of manufacturing that can cause budget expenditures to go out of whack and using data to control them.

Nonconformance Management

In the manufacturing processes, there will, of course, be some variance in the quality of production; this is unavoidable. The manufacturing manager must therefore understand the processes that are needed to manufacture (with the total amount of resources used), the fallout of those processes, and the resources needed to determine whether the units that have fallen out of production are scrap or recoverable.

Although the process of scrapping production units may be understood (key words being "may be"), the process for determining whether a unit should be scrapped is not always as clear. Depending on the industry (there is considerable variation per industry), there may be multiple attempts to recover a production unit (test, analyze, repair, or rework) before final disposition (finally scrapping or returning it to production), and in many cases, this time spent in limbo is not tracked very well. In the electronics manufacturing industry, for example, a production unit will be identified as nonconforming and then sent to a repair operation. The repair operator will analyze the problem to determine the cause of the nonconformance and then attempt to repair the unit. However, if the nonconformance occurs frequently, the cause may already be known and the operator can move straight to repair/rework. After the unit has been repaired/reworked, it must be retested (or inspected, depending on the nature of the problem). If the repair/rework was not successful, this process starts over again.

This kind of repeated *analyze/repair/test* process is common in many discrete manufacturing companies and is frequently referred to as the *hidden factory* because of the lack of visibility and process tracking in this area. More on this in later chapters.

The influence of the hidden factory can be felt in both direct and overhead labor, and although some managers try to make a rough estimate of these labor inputs, they can be deceptively hard to track and can vary widely. If the manager works closely with the CoE, the cost of the hidden factory can be better managed and accounted for in budgeting.

Continuous Improvement Management: An Introduction

Another area of manufacturing that is not well understood from a cost perspective is CI, which includes improving the product and the process of manufacturing the product. In some cases, it is necessary to update the product to improve the process of making it. If Product Engineering is considering a release of a new version of a product, the timing of that release is important. Frequently, as Product Engineering is waiting to release a new version of a product, manufacturing has to contend with implementing rework on the production units as part of a temporary deviation to the manufacturing process. Although there is a cost to implementing the release of a new version of a product, care should be taken to also consider the cost to manufacturing to maintain the temporary process deviation.

Real-Life Experience

During my time as a program leader for an ISO-9000 initiative, I was documenting the manufacturing process for a product line and found the line was using a temporary process deviation to manage some rework required to fix a design defect. When I asked how long this deviation was going to last, the line supervisor said that the deviation had been continuing for a couple of months, and Product Engineering had not identified when the new version was going to be released.

When I asked Engineering staff about the timeline, they indicated that no version release was currently planned for the product, and because it cost about $5,000 per release, they were not going to plan a release anytime soon. I then explained that it took three people to perform the rework to keep up the production rates and that it cost production more than $5,000 per month to perform the rework. Engineering had not considered the cost of manufacturing in their analysis.

The release was then quickly processed, and the new version of the product was made available a couple of weeks later. Engineering was very careful about delaying revisions from that point on.

In this scenario, Engineering had not considered the cumulative cost of manufacturing versus the one-time cost of a change in design. This situation highlights the importance of a formal CI program. The longer issues are allowed to remain in the production process, the more that cost accumulates. Within a single line (or production process), the cumulative cost of a problem can reach thousands of dollars in a short period.

In many cases, process changes can be implemented without a product revision change. The issue is understanding that many concerns must be evaluated when looking at product or process inefficiencies. Having Product Engineering working regularly with the CoE to evaluate the benefits of changes to the product as well as the impact of those changes on production can save a lot of money.

Expanding on the responsibilities of the CoE, as its team members gain more knowledge of the overall processes and a detailed understanding of the production fallout (including the characteristics of that fallout), the CoE can now take a prominent role in analyzing issues and managing the CI program, including analysis of the hidden factory. Having fully reviewed the processes and the characteristics of process fallout, the CoE can now provide much more accurate information to operations management for determining budgets and identifying where within manufacturing operations costs are starting to deviate from the budgeted estimates. An additional outcome of deep analysis of process fallout characteristics is a possible understanding of common issues across multiple processes. This not only allows Operations to recognize potential problems in multiple processes, but it also allows the CI program to present a much more coordinated response to similar problems that may have common corrections and that can be implemented across the board. An expanded discussion of the CoE's influence on continuous improvement is presented in Chapter 3.

Lost Production Units

Another issue that often arises in discrete manufacturing (more often than managers want to admit) is the loss of production units during various analysis activities. These activities can result from researching nonconformances within the hidden factory, but they can also happen as a result of the following:

- Quality taking units off the production line to analyze product characteristics

- Product Engineering taking units to analyze design characteristics

- Manufacturing Engineering (or Quality Engineering) quarantining production units for process or quality evaluations

In many cases, the removal of these units has not been scheduled (and is, therefore, an unexpected variance), and the expectation is that the production units will be returned to the production line "in short order." The problem is that as production units are removed from the production line without tracking controls, the temporary owners of the production units often forget to return the units until a much later date (if at all). Therefore, it is recommended that someone (a group that oversees production processes) act as the manager of the unit extraction from production. This is an area where the CoE can be of value. As the responsibilities of a CoE mature, the CoE members also develop relationships with key people in Product, Quality, or Manufacturing Engineering. As a result, the CoE can be well-positioned to manage these activities. If there is a need to remove production units from the floor for a longer-term (longer than the production run that they are "borrowed from"), it is best practice for these types of requests for units to be independently scheduled through production planning and the labor costs tracked and possibly accounted for in the budgets of other departments.

An issue similar to lost production units occurs with complex manufacturing (e.g., truck trailers, tractors, and other farming equipment) when the end product has multiple subassemblies that are being built. If there is a change in the requirements or design of the end product, the subassemblies currently in production may not match the updated end-product requirements. In many cases, the current subassembly will be "put aside," and a new subassembly will be started immediately in an effort to maintain the delivery schedule. The unused subassembly is usually moved to the "warehouse" until someone can determine how it can be either disassembled or reworked to make it usable. The problem is that no one ever gets to these units, and they continue to build up in what is commonly called the *bone pile*. These subassembly units will eventually be scrapped to make room for something else, but the details of what went into each unit and possible recovery of material or reuse in another end product are then lost.

Real-Life Experience

As an MES consultant, I was working with a company that made flatbed trailers. While mapping out the company's processes, I started to dive into its nonconformance processes to better understand how it dispositioned production units (keeping in mind that the company's subassemblies were large axle systems, flatbed frames, and full wheel assemblies). I recommended that keeping track of where they were in the dispositioning process would be very important in at least determining the company's scrap levels. After a review with the plant manufacturing manager, I was informed that the company did not have a "scrap" process. If a unit could not be used for the production order for which it was planned, it was sent to a special warehouse location where it would be disassembled and the parts recovered. I was impressed!

Interested in looking deeper into this process (I wanted to map it into the MES to automatically update the inventory of recovered material), I asked to look at the warehouse location and observe the process. After a long walk through multiple plants (the full facility covered about six football fields), company representatives and I finally arrived at a large field at the very back of the facility (the field was a bit bigger than a football field by itself). When I asked how they kept track of what was out there, the forklift driver who moved the material to the field said there was no tracking system. Anything that was recovered was brought back because someone remembered that a unit was recently moved out there. Further investigation revealed that most of the subassemblies in the field were years old and had rusted beyond use. By my estimation of the density of subassemblies and assumption of the cost per unit, I calculated roughly more than $50 million of subassemblies at this "warehouse location," and from what I could tell, 90% of the units had never been used and apparently none of the amount was written off.

Getting to the Data

Understanding how to use data is one thing; having the knowledge base and presenting data in the form of usable information is another. Maintaining the knowledge base and determining how to present data as usable information will become part of the CoE's responsibility. An additional concern for the CoE will be understanding how to get to that data. As mentioned in Chapter 1, an important step in the right direction is to initially establish a means of collecting some data. (Nike's motto, "Just Do It" comes to mind.) However, when using data becomes a major part of the operations mindset, it then becomes important to make that data collection and reporting efficient. Otherwise, the increased cost of gaining information from the reports (too much labor needed to process the data) and taking too much time to derive that information leads to the problem of aged data. Ensuring a common and efficient manner of data collection and reporting is another part of the CoE's responsibility—and this is where having a system like an MES can be important. An MES was designed to handle collecting data during production and reporting that data in real time and with the scope of visibility to the entire production floor. Whether the system is an MES or some other automated system of data collection is not as important as ensuring that the collection process is manageable and the data collected is easily accessible and interpreted for reporting.

By using an MES to collect data during production activities, the manufacturing equipment performance data can be linked to the production unit to compare with test and inspection results. Frequently, a relationship can be identified between a repeating nonconformance of products during a test and the drift in manufacturing equipment performance. This is the reason for creating the interfaces between the programmable logic controller (PLC)-level databases for manufacturing equipment (known as *operational technology*—OT) and the relational databases of enterprise systems (known as *information technology*—IT). As technology on both the OT side and the IT side improves, making it easier for these technologies to communicate and interact (called *IT/OT convergence*), collecting this data and linking it together becomes simpler and more efficient. (Refer to Chapter 6 for a detailed discussion on interfacing between OT and IT systems.)

What Is IT/OT Convergence?

Some time ago, we saw the start of automation in manufacturing. The automated equipment was designed to perform a specific function at either a speed or power level that humans could not achieve or to maintain quality in repetitive work. But when the automated equipment failed, the controls engineer would need to get into the system to determine what went wrong. This required going into the registers of the PLC systems to look for clues of the failure. This was the start of OT data usage.

It was then determined that having some data prior to the failure happening was good for diagnostics (to see what led up to the failure). As a result, manufacturers started using PLC system historians to hold longer streams of data from sensors, heaters, and automation actuators (the use of large amounts of OT data).

After that, we learned that by monitoring the OT data from historians, the supervisory control and data acquisition (SCADA) functions could provide operators with insight into how the equipment was operating and when a failure was imminent. This gave operators (or supervisors) a chance to act before the failure occurred. It required the SCADA system to have access to the data from historians and to use proprietary communications like Modbus or Profibus to make that data available to operators via human-machine interfaces (HMIs). This further expanded the use of OT data, but the monitoring was exclusive to the performance of the equipment.

Through observing these data streams (gauges, operator screens, etc.), it was determined that by understanding more about the drift of data, one could recognize, via the SCADA systems, not only when equipment failure was imminent but also that the data could be used to analyze failures in product quality at test and inspection operations that may be several operations later in the process. This created a need to have OT data linked with IT data and resulted in creating complex interfaces from PLCs and SCADA systems to IT systems like MES, which enabled relational data of product failure to be linked to operational data of the equipment making the product. Some tried (and are still trying) to direct it to ERP instead, but ERP systems are not equipped to process and present that data fast enough and in enough detail in relation to other operational events during the manufacturing process. The issue with the interfaces to the MES is that they are highly custom and expensive, sometimes adding as much as 40% to 50% of the implementation cost to a project. The next step in improving data access would be to make PLC data more accessible to MES and other relational databases. When this need for access is combined with the lower cost of computing (with computer memory becoming cheaper and network communication becoming more stable—enter 5G networks), it makes sense to connect the PLC sensors and actuators directly to IT networks (IIoT) and make their data available to… "whoever needs it."

This discussion highlights the fact that making OT data more readily available to IT systems is simply the next step in a longer process of the technology evolution that brought us to Industry 4.0 and IT/OT convergence.

An Additional Value of a CoE

In the recent survey performed by Tech-Clarity titled "The Manufacturing Data Challenge" (mentioned earlier), a major concern for companies that were having some success with transitioning to data-driven management of manufacturing was "the IT/OT divide" or the differences between the IT groups that manage the IT databases of enterprise systems (like MES) and the controls engineering groups that manage the OT databases at the controls level (PLCs, distributed control systems—DCSs, and SCADA systems). According to the survey, these two groups have "radically different mindsets and objectives." When looking at the focus of each group (as reported from the Tech-Clarity survey), it is easy to understand why there is such a divide.

Controls engineering focuses on the following:

- The performance of individual machines and equipment

- Ensuring that equipment is available to keep processes running

- Ensuring that the physical product is ready to ship

- Maintaining a quick response to any issues that arise to get the equipment back into production

IT professionals focus on the following:

- The performance of the production floor or plant as a whole

- Coordinating all data and documents to ensure an understanding of the complete history of the product is available

- Seeing the data package of the product as an important part of the delivery of the product as a whole (not just the physical product)

- The availability of the product's data package relies on standardization, completeness, and efficiency of transactions about the past and future

According to the Tech-Clarity survey, the companies that have been successful in transitioning to a data-driven manufacturing management focus have made an effort to ensure that these two groups work together. By developing a CoE that includes both the IT group and the controls engineering group (they do not have to be in the same department), the company is providing a venue that allows these two groups to work through the differences and optimize the solutions put forward to benefit the scope of both groups.

Summary

For decades, MOM has been a combination of collecting data over a period, reviewing that data in the form of static reports to understand where the company has been going (for that period), and then implementing best practices and using more and more detailed process models to try to estimate where the company must go. Because there are a lot of "moving parts" in most manufacturing operations organizations, it has always been difficult to "stay on top of things." In recent years, there has been greater emphasis on managing manufacturing operations by using real-time data and analysis to gain better visibility of the details of events on the production floor and to better estimate the path forward for those production floors. Getting the transition to a data-driven management model right is critical for the success of a company. As illustrated by the Tech-Clarity survey, several factors are key to a company successfully transitioning. Those factors include the following:

- A unified understanding of the interpretation of data end-to-end of the company

- Rapid access, distribution, and interpretation of data systematically throughout the company

- Actively establishing and managing standards in process, data collection, and reporting

- Actively managing (and standardizing) the processes used not only in manufacturing but also in the business as a whole, including management procedures

In the following chapters, I address the value of the CoE in many areas of manufacturing and the systems that can be used to help in the transition to a data-driven company.

3

Process Management, Continuous Improvement, and the CoE

Imagine that you have been hired by a discrete manufacturing company as a manufacturing engineer. Your first job is to define the process that is required to manufacture a new product. Creating (or documenting) a process covers a wide range of activities and includes things such as the sequence of steps that are required to make the product and the material that will be processed at each step. In other words, it covers how to build whatever subassemblies are required, how to attach those subassemblies or raw components to the main product, or how to use some of the bulk material as part of the processing. Documenting the process also includes defining what tools and equipment will be used at each step. As the manufacturing engineer, you have figured this out and understand the entire process of manufacturing the product. Now you are ready to start defining a production line . . . or are you?

One of the primary factors that makes manufacturing difficult to manage is *variation*. But why is variation such a problem? In the next section, I present a very simple product, establish a simple process to manufacture it, and then dive into the many things that can cause variation within that process. I invite you, the reader, to review the processes that I have defined and look for issues that may be of concern for an actual manufacturing floor. Because there will always be some aspect of variation, effort must be made to determine whether specific kinds of variation are significant or not. When implementing a continuous improvement (CI) initiative, a great deal of effort must go into analysis to isolate the effects of each aspect of variation and determine whether the specific cause of variation is a concern. In this chapter, I discuss the aspects of manufacturing processes that can cause variation and provide an example of a controlled methodology for CI. I also discuss additional responsibilities for the CoE.

Let's get started.

Product BOM (bill of material):

- 2 in × 2 in × 1 in deep plastic case (qty. = 1)

- 2 in × 2 in × 1/4 in plastic lid (qty. = 1)

- 2 in × 2 in rubber grommet (1/8 in thick) (qty. = 1)

- #6 × 1/2 in Phillips wood screw (flat-head, self-tapping, zinc) (qty. = 4)

Tools:

- #2 Phillips screwdriver

Assembly Process:

1. Place the plastic case in the work area (open side facing up).

2. Place the rubber grommet on the edge of the plastic case (matching the edges of the case).

3. Place the plastic lid on top of the rubber grommet (aligning the screw holes of the lid with the screw holes of the case).

4. Using a Phillips screwdriver, manually insert and screw down each of the screws until tight to fasten the lid to the case.

5. Place the completed case in a holding bin next to the work area.

The flow diagram of the process is shown in Figure 3-1.

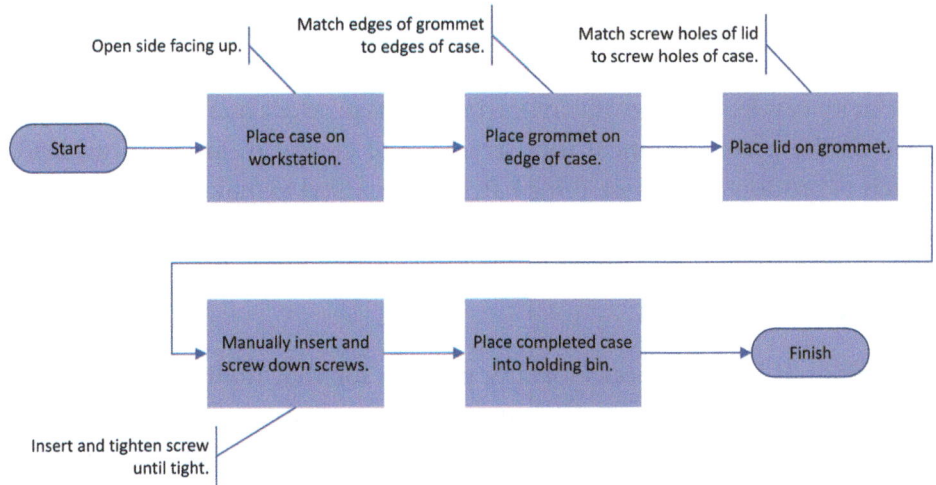

Figure 3-1. Simple process flow diagram.

One would not be faulted in thinking that the preceding instructions are pretty thorough and could be used to manufacture consistently.

However, one of the first issues that would arise relates to the instruction "screw down each of the screws until tight." Just how tight is *tight*? This subjective term can mean many different things to production operators, and how tight the screws actually need to be, depends on the product's application (where and how the case will be used). For example, if the instructions are for an electrical box for use indoors, just having the lid *snug* (which is another subjective term) may be enough. However, if this box is to be used underwater and is expected to be airtight, the "how tight" expression must be a lot more descriptive.

If the case is being used underwater, there would need to be a detailed expression of "how tight" which is usually a measurement of the amount of torque that is applied to the screws. To provide this instruction, the tool required is a torque driver and not just a Phillips screwdriver. With the instructions for *using* the torque driver, there is an additional process requirement to set the torque driver to the correct setting. But how accurate is the setting? When you set the driver to 0.2 ft·lb, for example, is that setting 0.2 ft·lb or 0.20 ft·lb (defining the resolution of the setting)? Another issue is that the torque driver may have the resolution in the setting, but is the setting calibrated against a national standard? Therefore, when it is set to 0.20 ft·lb, is it accurate and not 0.25 ft·lb when compared to a national standard?

This was supposed to be a simple process, and the instructions seemed good. What is going on here?

This is just one example of the need for explicit instructions and why detail in process management for manufacturing is so important. The set of instructions provided to the manufacturing operators is usually referred to as a *standard operating procedure* (SOP), and the need for clarity in the SOP cannot be overstated. In discrete industries, a lack of clarity can lead to significant variations in product quality. In process industries, such as the pharmaceutical industry, lack of clarity in an SOP can lead to an entire batch of product being lost. (One batch can be used to make hundreds of sellable units and cost thousands of dollars.)

Process Management

When defining an initial process (or any process for that matter), it is important to recognize that *the process is going to change*, and there is no such thing as *the* optimal process. You can (and should) optimize it from its current state in relation to a set of goals (i.e., what you are trying to achieve via optimizing). The goals for the optimization will define the primary characteristics that you (as a manufacturing engineer) are trying to achieve and measure, such as repeatability, reduced cycle time, reduction in cost, and reduced nonconformance. You must also understand that while pursuing improvement in one characteristic, you may be trying to either hold or temporarily relinquish another characteristic; there are likely to be goals that will also fundamentally conflict with each other. In this context, a business decision must be made. Note that before you can decide on a goal to optimize, you must first understand the current characteristics that can be optimized and their current values.

Defining an Initial Process

As covered in Chapter 2, several items must be defined when creating and documenting the initial process. The first is the outcome of the process (in manufacturing, this is the product). What are the characteristics of "a good product" in terms of physical specifications and performance characteristics? While defining the process, ensuring that the process does not damage the product's performance is, of course, just as important.

This was a major factor in electronics manufacturing. It took several years of manufacturing with semiconductors to discover that manually handling products during manufacturing was causing static electric shocks (static) to those products, sometimes at such low levels of energy that the operator was completely unaware of anything happening. Although the product would function during testing, the small static shocks the product experienced during manufacturing significantly reduced the product's operational life. Static control and the processes to manage it on the manufacturing floor are now a primary concern for all electronics manufacturing companies. Many electronics manufacturing companies now have detailed procedures for managing static during the entire manufacturing process.

If a process already exists, the initial step is to document what is currently being performed (including all known variations). The goal at this point is simply to have the process performed consistently and eliminate any controllable variation. This means retraining all production operators to follow the specific details of defined processes and to understand the variations that have been removed. When retraining production operators, it is also vital to ensure that *they do not get too "hung up" on this being the right process*. They must understand that the currently defined process is only to see "what must change." The process will include any testing or inspection that is being performed to ensure "acceptance" of the product and the fallout at these operations. (At this point, it is acceptable just to understand the total percent of fallout of the first-pass yield.) You will need to document each step within the process, what information is needed at each point in the process to perform the operation, where the information is coming from (data used to generate the information and source of the data), and who is expected to provide that information. (If this is a Lean value stream analysis, the cycle times and wait times for each operation would also be expected.) At some time, it will be necessary to review the process steps that were removed for causing variation and determine if they were relevant and should be reintroduced in a controlled manner.

Now let's go back to the original process for our plastic case.

First, the intended use of the product is to hold some kind of sensor (in this case, the specifics are not needed) and to protect the sensor from a damp environment. (The need to be airtight is important, but the product is not expected to be submerged in water.)

Manufacturing Process

Preproduction – The Phillips screwdriver called for in the original process is replaced with a torque driver that has been calibrated to apply a pressure of 0.20 ft·lb (a random value for the purpose of this example) and fitted with a #2 Phillips driver bit (sometimes called a *star*-head) screwdriver bit. In other words, this is an electric screwdriver that will tighten a screw with the pressure of 0.20 ft·lb (a measure of rotational force) and then stop driving the screw any further.

Operation #1 (previous assembly operation redefined)

1. Place the plastic case in the work area (open side facing up).

2. Place the rubber grommet on the edge of the plastic case (aligning the edges of the case).

3. Place the plastic lid on top of the rubber grommet (aligning the screw holes of the lid with the screw holes of the case).

4. Manually insert a screw into a hole in the lid, and use the torque driver to tighten the screw until it starts clicking (has reached the correct tightness).

5. Repeat step 4 for the remaining three screws.

6. Place the completed case in a holding bin next to the work area.

Operations #2 (inspection) – To ensure that the case was assembled correctly, the inspector will verify that the production unit has four screws installed, that the case top has not been misaligned, and that the grommet has been placed properly between the lid and the case body.

Note: It is important for any inspection (or test) operations to have specific points to check. Giving an inspector the function of "ensure it is assembled right" is just as subjective (and problematic) as the original statement of a "tight screw." As the process improves over time, the inspection criteria must be changed to ensure the parameters of the latest issue are being investigated.

1. Pick up a production unit from the holding bin of the previous operation.

2. Visually verify that four Phillips screws have been installed and that the screw heads are flush with the surface of the lid to the casing (the assumption being that if the screwheads are flush with the lid's surface, the correct amount of torque has been used).

3. Visually inspect the positioning of the grommet between the case and the lid.

   o Observe that there are no kinks in the grommet and that there are no gaps between both the grommet and the case and the grommet and the lid.

4. If the inspection conditions are met, place the production unit into the outgoing holding bin for packaging.

The detailed SOP is shown in Figure 3-2.

For this initial production process, the actual steps for the production operators (both assembler and inspector) have now been defined.

Now that the manufacturing process is defined, you can get the production line to start producing. After a few production runs, you (as the manufacturing engineer) talk with both the assembler and the inspector to see how things are going. The assembler acknowledges that things are going well. However, the inspector mentions a few possible issues that were easily fixed but were slowing down the inspection a bit. You ask for more detail about the rate of issues, and the inspector's response is a little vague. This provides the next insight into a flaw in your process. *As you have not provided a means of recording these issues, it is hard for the inspector to give feedback.*

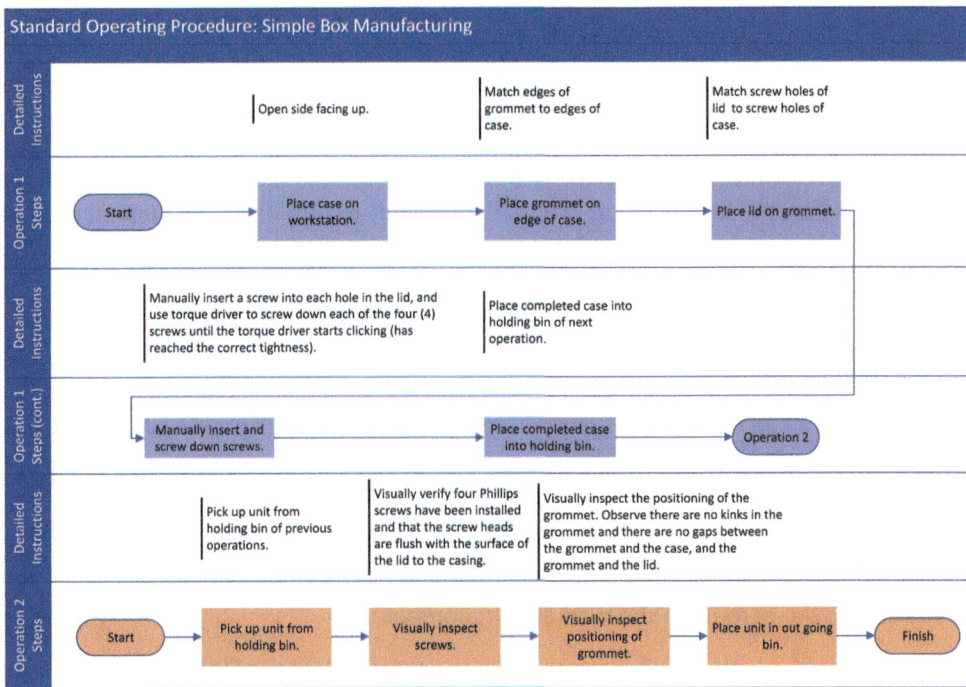

Figure 3-2. SOP (simple box manufacturing).

Adding Data Collection

Just as critical as which data to collect is *how* to collect the data. The most popular method of data recording is still the spreadsheet. This method is inexpensive and relatively easy to set up. The application manages its file structure, and the data structure that is used to record the data can be changed fairly easily. The problem with this method is that because it is isolated only to the workstation that is set up with the application, it does not easily relate the data from one operation to the data from another operation. In addition, the file size limits the scope of data that can be analyzed at any one time. However, for manufacturing environments that have a limited need for data collection, this may not be a problem.

In the case of the previous process, because the requirement is only to have visibility of the fallout rate, the spreadsheet option would be satisfactory. So, you set up a computer at the inspection operation and create a screen that tracks either a Pass or a Fail. If a Fail is recorded, a drop-down list is provided to record what was wrong. After talking to the inspector about the problems that have been encountered, you update the list to include other reasons for failure.

Now that there is a way of collecting data, the inspection process must be updated.

Operations #2 (inspection) – To ensure that the case was assembled correctly, the inspector will verify that the production unit has four screws installed, that the case top has not been misaligned, and that the grommet has been placed properly between the lid and the case body (this part has not changed, but the data collection must be added). Using the workstation computer, log on to the computer and start the spreadsheet application by double-clicking on the file shortcut. (You have made it easier for the inspection operator by creating a shortcut on the computer desktop that references the specific file needed.)

1. Pick up a production unit from the holding bin of the previous operation.

2. Visually verify that four Phillips screws have been installed and the screw heads are flush with the surface of the lid to the casing (the assumption being that if the screw-heads are flush with the lid's surface, the correct amount of torque has been used).

3. Visually inspect the positioning of the grommet between the case and the lid.

   o Observe that there are no kinks in the grommet and that there are no gaps between both the grommet and the case and the grommet and the lid.

4. Access the screen on the workstation computer.

   o If the production unit has passed inspection, select the PASS button.

   o If the production unit has failed inspection, select the FAIL button and then select a reason for the failure from the dropdown menu.

   o If the reason for failure is not available, select OTHER and enter a reason into the dialog box.

5. When the inspection is complete, place the production unit in the outgoing holding bin for packaging.

What happens to the production units that have failed? They can't go to packaging!

Therefore, the final step has conditional options:

6. When the inspection is complete:

   o If the production unit passes, place the production unit into the outgoing holding bin for packaging.

   o If the production unit fails, place it in a return bin to be sent to the previous operation to be fixed.

Now we must update the assembler's process to deal with the returns. After this effort is complete, you must train (and often retrain) the production operators to follow the defined process.

In an actual production environment, these failures (or nonconformances) must be dealt with, and the details at inspection must include identifying not only *what* the nonconformance is but *where* it is on the product and which production unit had the nonconformance (i.e., the batch or serial number). In addition, the instructions must guide the repair operator on how to analyze the cause of each nonconformance and how to correct it. However, because those steps are usually handled by trained repair operators, the instructions are more likely included as part of the training than as part of a process document.

The main point of this exercise is to recognize that even for a simple product, documenting the required details of the process to provide consistent manufacturing can take a lot of effort. When a production floor makes multiple products and/or also makes subassemblies that are used later in the process of making multiple end products, documenting (or creating) the entire process can be quite complex. Following these instructions consistently is just as complex. Because production operators must be able to move easily from one production line to another, using consistent terminology between SOPs is important. This is one of the primary reasons for establishing a CoE that has a diverse set of skills and maintains a centralized knowledge base for MOM.

Redefining Current State

There are two primary reasons for redefining a process.

They are:

1. **Investigation** – Temporarily changing the parameters of a process to gain more data about a specific issue.

2. **Improvement** – Permanently changing the parameters of a process to improve the process's capability according to a specific goal or to implement *the* solution to a specific problem.

   o Recognize that *permanent* is a relative term and is only valid until the next change in process is required as part of ongoing CI.

There are also different aspects of investigation:

- General characterization of a process to determine "what issues are."

  o This may require less specific data collection but a wider distribution of data collection over the entire process.

- Characterization of an explicit issue to determine the occurrence rate and detailed characteristics of that issue or the characteristics of one or more possible solutions.

o   This may require considerable detail of data being collected at a single point in the process.

Understanding these differences can affect which data will be collected and which process changes are needed.

In our simple process of assembling the plastic case, adding the inspection and data collection is an example of general characterization. Part of the scope of this type of investigation is to determine the failure rate for each type of nonconformance. By identifying the types of nonconformances and the quantity of each type (determined by counting the number of times each nonconformance was selected), you (the manufacturing engineer) are in a better position to characterize the fallout from the first-pass yield (FPY).

After implementing the data collection in the simple process and collecting data for a couple of production runs, it is time to analyze it. You go to the workstation's files, copy the current file, and create a new file (same data structure but without the data collected) to ensure that the next analysis contains only new data.

In many cases, to analyze data from a general context, the first step (as mentioned earlier) is to determine the count of each of the different types of nonconformances. In data analytics, this is referred to as *clustering*. A typical reporting format for the output of clustering is a Pareto chart. Figure 3-3 shows an example of a clustering output presented in a Pareto chart. Assume that this is the output of your first analysis from the spreadsheet.

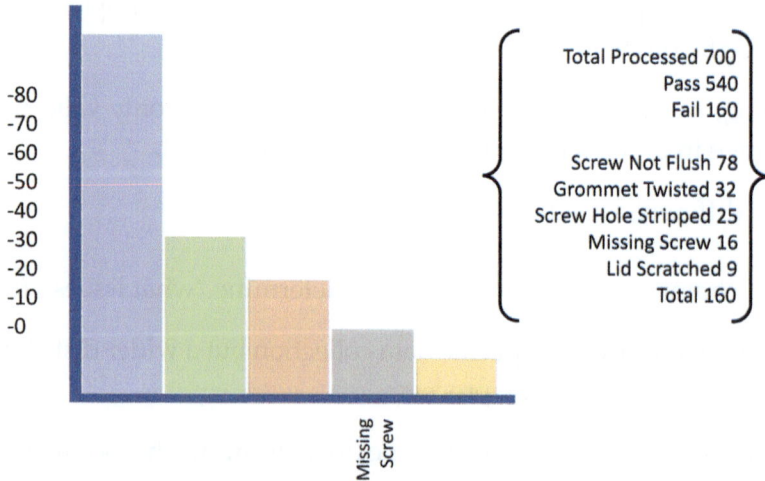

Figure 3-3. Pareto chart of clustering output.

With this data, the report shows that the process has about a 23% fallout rate, with "Screw not flush" as the most prominent issue, accounting for about 49% of the failures. With this process characterization, you can now say that you (more or less) have a defined process.

With the initial analysis complete, it is time to transition to more of a continuous improvement mode of operation. The problem with the current analysis is that it brings up more questions than it answers. That is usually the case with general process characterizations.

Now back to your work as a manufacturing engineer.

The Process of Continuous Improvement

Many companies will take on a continuous improvement initiative as a result of a customer complaint and will measure the initiative's success as either the money that has been saved (because the process is more efficient) or (more likely) the customer saying they are now happy with the company's output. There is, of course, nothing wrong with making your customer happy, but the problem with this mode of selecting continuous improvement initiatives is that it is purely reactive. With that mode of selection also comes the problem that some time may have passed before the customer complained (i.e., there was a period of time with a negative impact on your business), and the customer that complained may not be the only one that sees the problem (meaning you have been getting a bad rap to your reputation for some time from multiple customers). We are taught in operations management courses that approximately one in eight people complain about a problem, meaning that every complaint you receive represents seven additional customers experiencing the same problem but not reporting it. It also means that eight people are potentially unhappy with you, which could be causing a loss of business and reputational damage.

Another issue with selecting initiatives via customer complaints is that the focus is then on correcting the fault they have complained about. This focus (which may be good for the customer) provides no guidance on reducing cost unless the complaint is about the cost. If cost is the customer's concern, they likely are already buying from your competition instead.

Having a robust continuous improvement process helps keep your customers from going to the competition. Consider the following example.

Real-Life Experience

While I was working in a contract manufacturing company's MES team (the initial stage of a CoE), it was brought to our attention that one of our customers was on the verge of pulling its product line from the company. Our company had three plants that were manufacturing a specific product line, but the processes for manufacturing were not only different from each other but they also had different performance capabilities and reporting. From the customer's perspective, dealing with each plant was like dealing with a completely different manufacturing company. The customer's frustration grew as problems that were resolved in one plant would continue for weeks at a different plant unless the customer's engineering team spoke up. This led to concerns from the customer that the plants were simply not talking to each other (which they were not). The department I worked for was brought in to stabilize the use of the MES. It was during our review of the processes that we challenged the plants to resolve their process differences, which resulted in the initiation of a common process management group to support that customer. It took a lot of effort, but it worked out in the long run. The customer not only kept the product line with the company, but having gained confidence in our process management, they also switched other product lines to our plants.

Managing the Continuous Improvement Process

One of the first aspects of managing the continuous improvement process is to understand that there are generally four stages:

1. **Detection** – Creating the test or inspection capabilities that enable detecting variances and nonconformances

2. **Correction** – Analyzing the steps or conditions needed to correct the occurrence of variances or nonconformances

3. **Prediction** – Creating the capability (updating equipment, tests, data collection, and analysis) to predict when a variance or nonconformance is *going to happen*

4. **Prevention** – Changing the process, equipment parameters, or operator SOP needed to prevent the variance or nonconformance from happening

To ensure that processes actually improve, all four stages must be completed. The issue that many production floors have is that they accept that fallout is a normal part of the production process and, as a result, do not always complete all four stages.

In some cases, just detecting a nonconformance and repairing it or having the process create an alarm when the conditions are trending toward a problem (prediction) is very common. As long as the product is repaired before shipping or an operator can adjust the process to correct the trend, many companies are content to stay at the early stages of continuous improvement (correction) but will continue to suffer the cost of supporting the repair process (remember the hidden factory?) and occasionally miss correcting the problem.

As mentioned in previous chapters, a significant responsibility of operations management is to develop a budget for the annual plan (both for the cost of manufacturing for the product volumes estimated by Sales and for the overhead needed by the initiatives to maintain and improve the processes) and to maintain operations within that budget. The goal of operations is to produce by the most effective means possible and to deliver products "on time" (according to customer demand) and at the lowest cost. There is also the operational concern of recognizing that, in some cases, the cost of completing all four stages of continuous improvement may be prohibitive, and intervening when a trend is detected may be the best achievable solution.

Because, as stated in Chapter 1, variance in the process causes variance in cost, priority is given to reducing variance by every quality management methodology used in manufacturing. The key to reducing variance is standardization. But standardizing the manufacturing processes is not the only issue.

As mentioned in Chapter 2, it is important to improve not only the manufacturing processes but also the process of improvement. This requires documenting the manufacturing processes that drive production as well as the business processes that drive CI. One way to do this is to measure the cost of implementing an improvement (analysis of the problem, determining a solution, and implementing the correct solution) and compare it to the money saved after the improvement is implemented. However, because each continuous improvement initiative will be different, this comparison is not going to be easy. A familiar tool is the Pareto chart that was introduced earlier in this chapter. The key benefit to using this type of chart is that it highlights concerns in an acceptable general priority. As stated, the issue with using a Pareto chart is that it is data intensive and is best used in an environment where data is easily accessible. As a result, many MES provide some reporting in a Pareto chart format. However, the complexities of data can also make a Pareto chart hard to interpret correctly. If getting a Pareto chart correct is complex, what makes it worth the effort?

Using Pareto Charts

Developed by economist Vilfredo Pareto and popularized in the quality industry by management consultant Joseph Juran, the Pareto principle states that 80% of issues affecting a situation are influenced by 20% of contributing factors. This principle can apply to many areas, from donations to charity to the drivers of economics. However, from a quality perspective, it means that 80% of the occurrences of problems affecting production can be traced back to 20% of the causes. To put this another way, if operations managers were to *identify and clear up that 20% of causes of problems, they would recognize an 80% improvement in performance.* As a result, from an investment perspective, it makes sense for operations managers to focus on identifying that 20% of causes to get the greatest return on operational investment.

Example of Analysis

Returning to the simple process example that was used earlier in the chapter, we can now look at continuing your work as the manufacturing engineer to improve on the process. When we left the manufacturing floor, you were interested in using the general numbers to characterize the process. The main reason for establishing the numbers in the first place is to ensure that for any step you take to improve the process, you first understand which problem is causing the most fallout (and presumably the greatest cost variance). After you determine which problem is the greatest concern, you then need more data collection and analysis to determine the root cause and possible solutions to that problem.

So, going back to the Pareto chart presented earlier in this chapter (Figure 3-4, a repeat of Figure 3-3), you see that the problem with the greatest occurrence is the issue of the screws not being flush with the lid. Because this problem accounts for 49% of the fallout, it makes sense to focus on it. The screws not being fully installed could be an

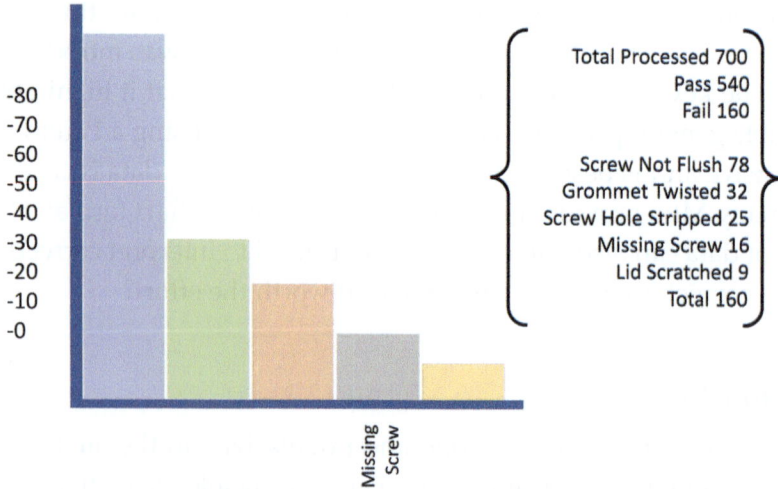

Figure 3-4. Pareto chart of the simple process.

indicator of a few issues. After performing an Ishikawa analysis (frequently referred to as a *fishbone diagram*), you determine the following possible root causes:

- Not enough torque is being applied to the screw from the torque driver. This can result from

  o the torque driver not being set and calibrated properly or

  o the plastic sleeves on the case for the screws being harder than expected (requiring more torque than the current setting).

- The operator is not applying the torque driver fully (until the driver starts clattering when released).

- The torque driver has variations in applying torque to the screws despite the calibration.

Possible root cause #2 may be easier to determine than cause #1 or #3. So, you look into this cause first.

While the line is engaged in a production run, you discreetly monitor the activity of the two operations to ensure that the assembly operator is, in fact, following the process and fully driving the screws into the lid of the case until the torque

> As long as there is considerable variation in the process, there is also going to be considerable variation in the results.

driver releases. (This is what causes the clattering sound when using a torque driver.) While observing, you do not see an issue with the assembly operator's activity, and when the production run is finished, you copy the data from the spreadsheet again. A quick analysis shows that the issue of screws not being flush is still there, but the numbers are considerably less than in previous runs. In a production run of 400 units, you now have 18% fallout, with 30% of the fallout caused by the screws not being fully inserted (the total count of "Screw not flush" is about 22). The first question that arises is, Why is there a difference? This highlights one of the more significant issues of process management: As long as there is considerable variation in the process, there is also going to be considerable variation in the results. This makes it hard to do the following:

- Identify the most significant problem, particularly if the results of a Pareto chart vary from production run to production run, as they will.

- Determine the root cause of the problem because too many possible causes "show up" in the counts and then seem to "disappear" in the next production run.

In process management, the more consistent a process is, the easier it is to improve.

For now, you decide to overlook the issue of variation in the Pareto chart because the issue of "Screw not flush" still has the highest occurrence. The next possible problem to consider is variation in the tool applying torque. For this, you set up an experiment that will measure the maximum torque applied before the tool causes the driver to release.

Conducting a test that uses a calibrated torque meter enables you to measure and save the peak torque that was applied to the meter until the meter is reset. For the test, you follow the directions you created to adjust the torque driver at the start of the shift and then apply the torque 20 times per set. You then reset the torque driver to zero and repeat the setup adjustment. After testing the applied torque for a total of 60 tests and recording each result, you are comfortable that you have a good characterization of the capability of the tool and you create a report on the results. (Technically, a good characterization would probably require 200 samples or more, but this is a good start.) Using the report, you can determine if the torque driver is delivering consistent results and if the driver is always within specification. However, first, you need to know the acceptable specification range. After determining whether the driver is consistently within range, you would then need to look at the consistency of manufacturing the plastic case as there may be variations in the density of the plastic sleeves for the screws.

In this example, you will need to investigate each of the possible causes until the root cause can be explicitly defined. After the root cause is defined, further action is required to determine a solution to the problem. Then you will need to go back to performing a general characterization to determine the next most significant issue in the manufacturing process. Managing the CI process requires investigating the acceptable requirements of the product being produced to ensure that it properly specifies those requirements and the steps being used during manufacturing to ensure that the currently defined specifications are being met.

This highlights the importance of managing the CI process in addition to managing the production process. It takes a few rounds of analysis to determine the specific root cause of an issue. Then it takes a few additional rounds of analysis to determine the correct solution. Using a Pareto chart to focus attention on the largest issues helps ensure that any initiative will provide the greatest benefit to operations. Another aspect of continuous improvement is coordinating activities across department lines. This helps to prevent redundant investigations and analysis efforts.

Managing the Process of Process Improvement

Most people with an operations background are familiar with the Deming cycle of Plan, Do, Check, Act shown in Figure 3-5. It is the basic model of the CI process and is a part of every quality and process improvement methodology (Lean, Six Sigma, Theory of Constraints, etc.) in one form or another and is fundamental to most analysis processes.

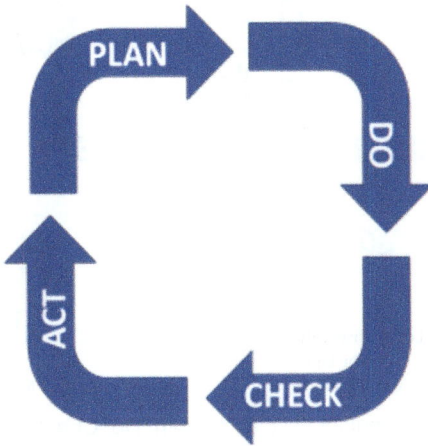

Figure 3-5. The Deming cycle of Plan, Do, Check, Act.

But is CI really that simple?

In most cases, it will take multiple iterations of the cycle to determine the root cause and even more iterations of the cycle, combined with a method of analysis called the design of experiment (DoE) which is used to isolate and test independent process parameters to determine the correct solution. The number of iterations of the Deming cycle depends on the quality of the data and the integrity of the analysis process.

Difficulty in collecting data means more Deming cycles will be required to fully analyze a problem, and a lack of process integrity (and failure to use DoE) could mean skewed (or blatantly wrong) analysis results. Both require a longer time to identify the root cause of a problem and the correct solution.

As shown in Figure 3-6, depending on the problem being analyzed and the number of people involved in the initiative, the cost of iterations can quickly add up, and this does not even include the cost of the original problem to Operations.

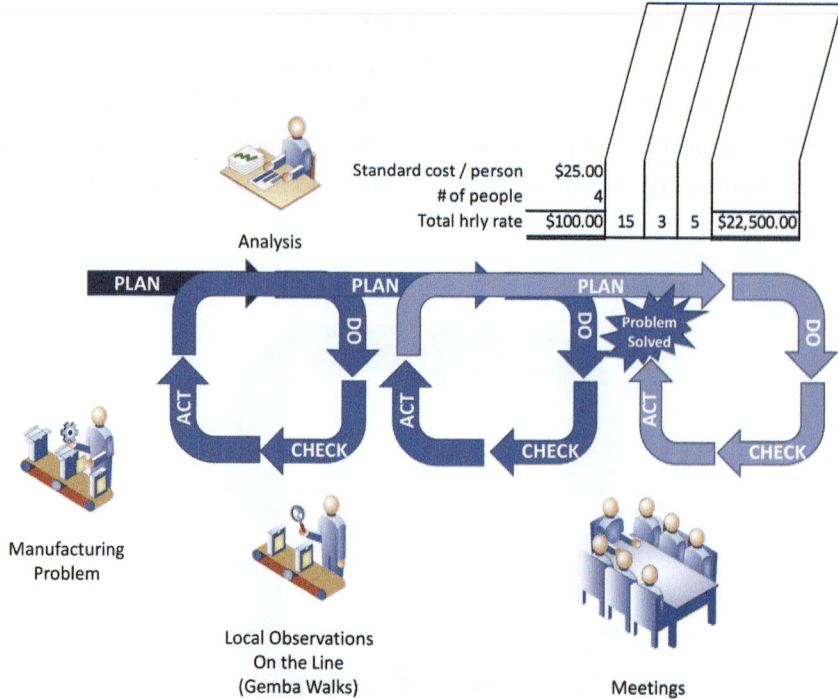

Figure 3-6. Example of the cost of multiple cycles of Plan, Do, Check, Act.

Figure 3-7 illustrates that the cost of analysis is even greater when each department investigates a problem, interprets the analysis, and creates a solution that is localized to its own scope of influence.

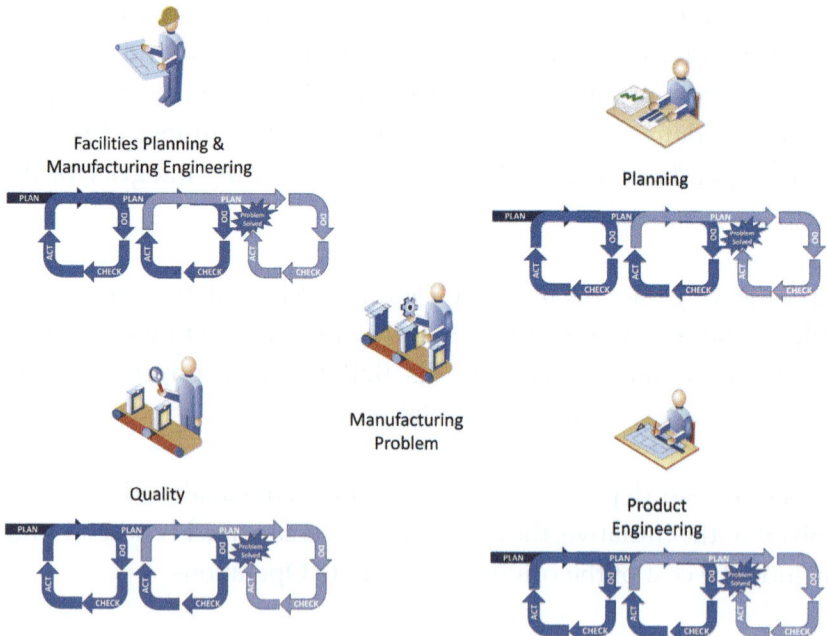

Figure 3-7. Multiple departments with Plan, Do, Check, Act.

The result is a significant amount of money invested in a limited analysis to fix a limited scope (or the incorrect scope) of the problem. When a company has a well-defined CI program and a centralized knowledge base for analysis results, these additional costs can be managed. As shown in Figure 3-8, it is also important to ensure that the interpretation of the analysis results is consistent and holistic to all departments that are stakeholders in the problem under investigation. This is an area where a diverse CoE can provide services to the entire manufacturing company.

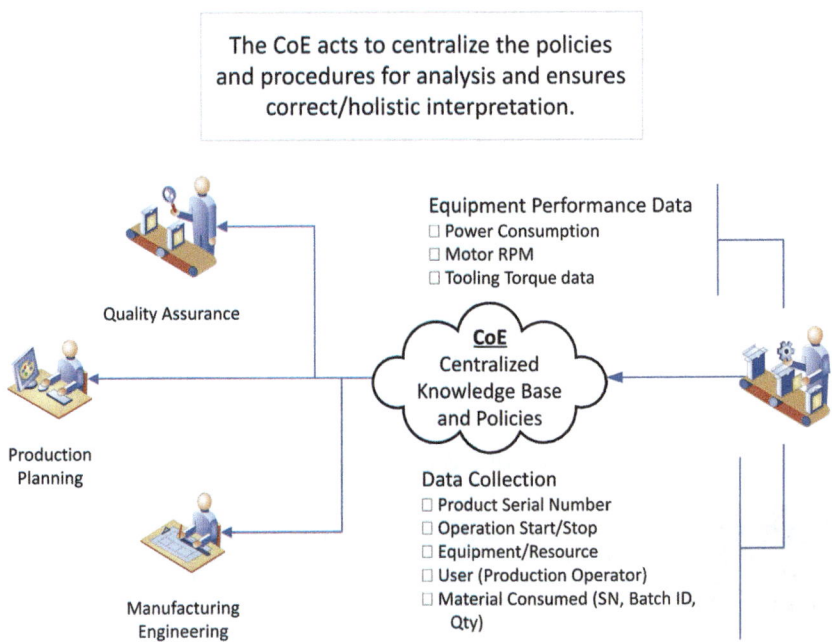

Figure 3-8. A CoE ensures a holistic interpretation of the analysis.

If you rely on the concept that repeatability reduces variation and reducing variation reduces cost, you can then generate guidance by documenting the CI process and working to make the *improvement process* repeatable.

This continued iteration of improvement is the foundation of the "golden batch" in the pharma industry. The initial runs of a production batch for a product will specify the "control limits" (a statistical process control parameter) of equipment settings for an acceptable production run (parameters needed for a "golden batch"). Analysis of the production run might reveal that the majority of the equipment parameter readings are within a narrower range than the current control limits. On subsequent runs, the control limits will be defined to be narrower, and steps will be taken to ensure that the narrower control limits can be maintained. The "golden batch" parameters are then adjusted for any subsequent production runs.

An additional driver of business management tools is the process capability maturity model (PCMM). The fundamental concept of PCMM is that there are five stages of process management maturity and by increasing the level of maturity in process management, a company is creating a management environment that provides the greatest capability to manage the continuous improvement program.

Process Capability Maturity Model: An Introduction

The PCMM is designed to be used in a company as a reference to understand where the company is with respect to maturity in process management and as a guideline to plan CI activities. Multiple industry organizations develop a specific definition of each level in the model, and each step in the model is intended to be a foundation for the next step in the model. In this book, I do not promote or recommend any particular definition of the model, but I do recommend adopting *a* model to be used consistently in a company. In this section, I provide a general description of PCMM and outline some characteristics of activities in a company at each level.

Figure 3-9 outlines the five levels of PCMM and provides a brief description of each level. The PCMM levels start at the bottom at Level 1 and build on capability up to Level 5.

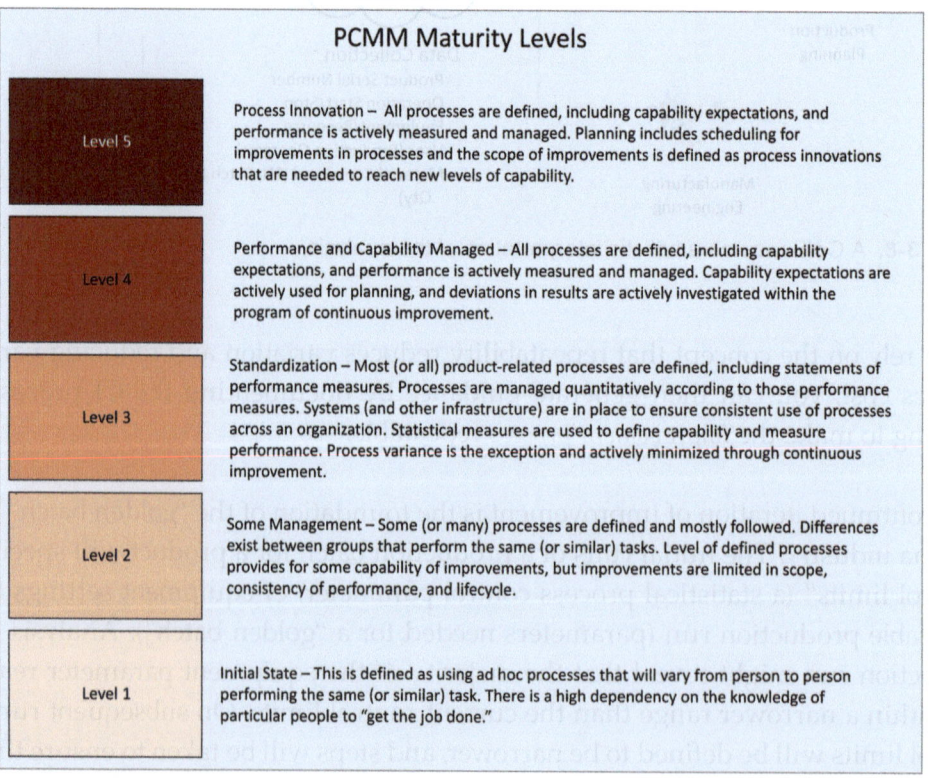

Figure 3-9. A PCMM model.

PCMM Level 1 – The first level of the model is recognized as the state of a company that has little to no process management. Processes (although they may be defined) are not necessarily followed. Each production operator performs an operation in their own manner, and with experience, that operator will modify their particular process according to their limited scope of the operation. This level of capability usually depends on the knowledge of certain people within the process and is driven by a management attitude of "just get the job done." Any activity for improvement is usually to resolve a customer complaint and is only recognized by the group that is directly impacted by the complaint. As a result, improvements are limited in scope, not transferable to other areas, and limited in life span because they are not carried forward as products and equipment are changed. Budgeting for this environment is usually overstated to protect against overruns, but adherence to the budget is purely dependent on luck. Frequently, management in this environment is more concerned about getting shipments out the door *at any cost* and will rely on key people to perform exceptional tasks, which often leads to those people burning out.

PCMM Level 2 – The second level of the model is recognized by isolated islands of processes being defined. When processes are defined and followed, each production area follows its own defined processes and seeks to improve process capability within its island of responsibility. This level of capability usually depends on localized knowledge at a team level and is frequently driven by a management attitude of competition between production areas. As a result, any activity for improvement is usually guarded by the group that drives the improvement initiative, and information is not shared. Similar to Level 1, improvements are limited in scope and are not easily transferable to other areas. Budgeting for this environment is still usually overstated to protect against overruns, but adherence to the budget can be a little more dependable. However, exceptions are frequent.

PCMM Level 3 – The third level of the model is recognized by the definition and use of processes across internal company boundaries. These processes are recognized as the company's current best practices and are applied by default to new scenarios. If there are situations where the best practice processes cannot be applied as is, action is taken to understand why changes are needed and a solution that is the least detrimental to the best practices is selected. Improvements are implemented via test scenarios to achieve the most stable results and verified at the time of implementing a new solution. The improved process is then implemented across the organization. At this level, there is a strong need for a central group to implement and manage a company's best practices (frequently the initial development of a CoE), and managing departments (or other plants) is more collaborative. Budgeting becomes more accurate, and adherence to the budget is more reliable.

PCMM Level 4 – The fourth level of the model is recognized as having processes that are "under control," and both capacity and performance measures are defined quantitatively. New processes that are defined are much more reliable from the start because the "process for defining a process" is managed and repeatable. Continuous improvement steps are more clearly planned, and proposed solutions are clearly defined and achievable. "Improvement of processes" applies to production processes as well as production support processes (maintenance and quality support processes). Management of best practices and continuous improvement is centralized and is overseen by senior management's sponsorship, and senior management significantly uses and relies on the output of this centralized group for all planning at the sales and operations planning level.

PCMM Level 5 – When a company reaches the fifth level of PCMM, most of its improvement activities are focused on improving (or achieving) strategic goals. This is a result of having a clear understanding of the processes that drive manufacturing from the production context and from the production support context (sometimes referred to as the *business side*), the planned capabilities and historic performance of those processes, and the issues that are limiting those processes (either in capacity or performance). These outstanding issues frequently exist as a result of external influences that the company cannot control or a planned decision not to invest in overcoming those issues and to use mitigation techniques instead. Reviewing all processes within the company is a regular activity actively sponsored and supported by senior management.

As stated in Chapter 2, the capability maturity model (CMM) started in the software development industry, and the PCMM model, in general, has become a major component of business process management (BPM), with some companies pursuing PCMM growth as a major part of their strategic planning. There are, however, many companies that never make it past the Level 2 capabilities as a result of a poor commitment to changing the culture of the company (particularly, senior management). If a company has multiple plants, the relationship between the plants is frequently a barrier to improving well into the Level 3 capability. Analysis of these companies by multiple BPM organizations has revealed that a lack of consistent management policy has been a significant factor and competitive attitudes between plants have added to these problems.

PCMM Program Ownership and the CoE

As discussed in the previous section, having an inconsistent set of policies for PCMM can make growth in process maturity nearly impossible. It can also make the program too costly to maintain, which will result in management pulling support for the program. A CoE can ensure that PCMM is consistently used throughout the company and that its importance continues to be recognized by senior management. By supporting a centralized approach to implementing process management, the company establishes clear ownership of the knowledge base (and, by virtue of ownership, ensures the distribution and use of the knowledge base). As mentioned, having the CoE as the owner of the process management knowledge base helps ensure that management policies are consistently applied. It also helps ensure that programs that rely on the knowledge base (continuous improvement, ISO-9000, Lean/Six Sigma) use it consistently. Ultimately, the purpose of using a process management model like PCMM is to improve the company. In any company, the right data will lead to information. That information will then lead to knowledge. Having a CoE own the knowledge base created by the PCMM ensures accountability for the development and distributed use of the knowledge base, and it promotes the success of the PCMM program.

Additional Thoughts

Although formal programs of continuous improvement (like Lean or Six Sigma) are beneficial to kick-start a company in the right direction, they are not required to implement continuous improvement. The primary requirement is to ensure that there is a methodical approach to analysis and improvement and that the methodology used is applied consistently across company boundaries. However, many companies have also found that developing (and implementing) their own methodologies and practices (instead of using a program like Lean) can make it more time-consuming to accomplish the same outcome than implementing Lean (for example) in the first place.

Good process management requires reliance on data. The key factors are knowing which operational data is needed for any particular issue and understanding how to represent operational data with the data collection capabilities that are available. When determining the data collection requirements for the issues you face, it is important to always reflect on the actual operational requirements. Ensuring the proper context of data and reporting is a significant function of a CoE in process management.

Summary

In this chapter, I took a detailed look at documenting and continuously improving a process. I also presented some industry models that manufacturing companies can use without having to "reinvent the wheel" for adopting best practices. This information provides a framework for implementing and managing CI. The key takeaways from this chapter, however, are the importance of the use and support of a centralized knowledge base of best practices and the support of a strong CoE program. Implementing a good CI program does not require a CoE, but the effectiveness of a CI program is greatly improved by also implementing a CoE.

4

Industry 4.0 and Data Mining

If you have been part of the manufacturing industry in recent years, you have probably heard the terms *Industry 4.0* and *data mining*. So far, I have discussed process management and quality management issues and have reflected on the CoE as part of the management program. I have also discussed the changes needed in manufacturing management to be more "data driven." But what is it about manufacturing that makes data-driven management so important, and why is Industry 4.0 such a significant part of that management?

Data Sets and Data Models

In the "Going from Symptom to Cause" section of this chapter, I discuss the difference between fixing an issue or implementing mitigation. It is always better to fix the root cause than it is to implement a mitigation strategy. However, there are times when a mitigation strategy is necessary. For example, mitigation might be called for when more investigation is required to determine the root cause and a control is needed in the meantime, or when it may simply be too costly to implement a root cause fix and a mitigation strategy is the only viable option. Whether it is part of a mitigation strategy or part of an investigation into a root cause, there is a need to know *when* something is happening.

Whether a nonconformance or a specific series of process characteristics occur, we need to be able to define what those characteristics look like. If the issue is physical, you can describe it in size, color, hardness, or some other physical aspect. However, when we want a computer system to find or detect something, we must be able to *describe what it looks like in data.* Describing an instance of something from a data perspective requires looking at what is called a *data set*. When there is a commonality to several concepts, and as a result, their data sets share common characteristics, defining the structure of those common characteristics within their data sets is called a *data model*. For example, each time an oven drifts lower in temperature, there is a starting temperature, an ending temperature, and a rate of change in temperature. This data for a specific incident of drift is a data set. If we wanted a computer to detect the drifting of the oven temperature, we might describe it as "a negative change in temperature of 5°C or more over a series of 10 readings (the number of readings may represent a time duration)," this description would be the data model for detecting that change.

Building data models is not new. Scientists, statisticians, and other professionals have been analyzing data sets stored on paper (the earliest form of a database) and creating data models (usually in some form of mathematical equation) for a long time, long before the digital age. What has changed is the use of computers and electronic databases to store large volumes of data sets and the ability to quickly analyze these data sets to create and validate complex data models.

As computing power is becoming relatively cheap and commonplace, more detailed data sets are increasingly accessible, and programming and using data models are becoming much easier. Through the field of data science, we learn more about the interactions between different activities via analyzing data sets and then creating (and understanding) the interactions via much more complex data models. But what does all this mean to a manufacturing plant?

Manufacturing Characteristics and Why Data Mining Works

There are some differences between manufacturing for the process industry and manufacturing for the discrete industry (*Ok… when you get into the details, there are a lot of differences*), but the characteristics that make data-driven management so important are common to almost all manufacturing industries. The first characteristic is that unintended variation is bad. Some might say that variation provides a kind of uniqueness to each product and may be aesthetically pleasing. However, in manufacturing, each of those *variations* is an opportunity for product failure, and because failed products do not sell and can damage a company's reputation, *variation should probably be avoided.* In extreme cases, the inability to control variation can cause a company to go out of business.

In defining a production line, many production steps are broken down into small segments that are each performed by a different operator or machine. When you remove variation, you get a highly repetitive set of operations that combine to, hopefully, build the correct product exactly the same as all the previously built products and you minimize the likelihood of nonconformance. If there are manual steps in the process, the goal of the manufacturing engineer is (or should be) to cause the operators to develop muscle memory for these highly repetitive steps. This makes it easier for the operator to detect if something is different and possibly out of place. A good production operator will recognize the importance of this muscle memory (even if it is to allow them a chance to talk to others while on the line without making mistakes.) But if most manufacturing engineers understand this, where does the data relationship come in?

When all the operations and steps in the manufacturing process are mapped out, the result is a highly sequenced instruction that defines not only the process for production to follow but also all decision points, the criteria for making decisions, and a definition of the outcome of each step in the process. It is a "model" of the physical activity needed to successfully produce a product. The problem is that unless we have the entire process videotaped, we are not able to determine to what detail the process was followed. We then turn to computer systems. As we look for information technology (IT) and operational technology (OT) systems to replicate this physical model (both for process control and for data collection), it is important to be able to *define the physical model as a data model*.

With today's technology, there are more and more ways to have sensors that can provide data that will then characterize each step in the process in greater detail. When all this detail is mapped out for each product's serial number, we can see in the reporting (depending on how we visualize the data) the occurrence of small variations. For example, the reporting can show when a production operator got distracted for a moment and did not quite follow the process, or it can show when a machine that has been running for a long time is starting to drift in its repetition as a result of wear.

Recognizing that there are always going to be small variations that happen during a process, the manufacturing engineer must be able to determine whether these variations resulted in a nonconformance of some sort because many of these variations will not be substantial.

> If there are a lot of small variations, it becomes harder to determine *which* variations are significant.

Hopefully, there is a limited number of variations that are significant in the process, but this will depend partly on the plant's maturity in process management. And it is important for the manufacturing engineer to remain focused on the significant variations that are causing the most grief (remember the Pareto principle). If there are a lot of small variations, it becomes harder to determine *which* variations are significant, and as the focus of improvement moves to increasingly subtler variations, a combination of issues can be a problem. As an abstract example, assume we are measuring an oven temperature. A manufacturing engineer may find that variation of parameter x (actual temperature) of up to $\pm10°C$ may be acceptable. However, when there is a drop in the temperature of more than $5°$ in parameter x *and* there is a variation in the speed of the conveyor feeding the oven that speeds up the conveyor by greater than two links per second, this combination may cause too great of a fluctuation in heating during a material's curing process and therefore create a nonconformance. With production running at full speed, how is a manufacturing engineer supposed to find a problem like this? This is where machine learning and other artificial intelligence (AI) concepts come into play and why it is important to maintain a relational database that can store a holistic view of the end-to-end process.

An Introduction to Data Sets and Data Models with MES

In the general configuration of an MES application, the system will provide the correct process that the product is expected to follow, including the type of equipment to use, the material to consume, and data to collect about the equipment's operational parameters (and when within the process to collect these parameters). As part of the MES data structure, the expected process, the product structure, and the acceptable criteria for all decision points will be defined. These master data elements (also called a *recipe* in some manufacturing models) will be used to guide production in the sequence of steps that the manufacturing engineer has provided to replicate the physical model that was discussed earlier. While the product is being manufactured, the MES will record all activity performed on the product, record equipment parameters while the equipment is acting on the product, and link this data back to the individual production unit (identified by a serial number or a batch number). During any test or inspection operations, the MES will also record the results of the test/inspection and any nonconformances (again, with a link back to the individual production unit). After all this data has been collected and stored with similar relationships back to the production unit, the outcome is a data set of all activity performed on the product and the operational parameters of the equipment used on the product during production. When this data set of activity (also called a production unit's *complete build history*) is compared with other data sets of similar production units (or a product's data model), a tremendous amount of information can be derived regarding variations in the production process. This is what makes an MES so powerful in implementing data mining and machine learning.

Although the general function of any MES supports this kind of analysis, there is a difference between the structure of this data as it relates to the MES supporting discrete manufacturing and the MES supporting process manufacturing. This difference will be examined in Chapter 6.

Using ISA-95 Data Models

In the ISA-95 set of standards, data models are introduced in Part 1, "Models and Terminology," and described in detail in Part 2, "Objects and Attributes." It is important to recognize the difference between the ISA-95 operational data models and the IT data models used in the schemas for MES, for example. Many MES vendors have used the ISA-95 operational models to define the general structure of the schemas. However, the data structures for MES must handle more detail than the ISA-95 models account for; this is to be expected as the MES must virtually replicate a particular production floor, whereas the ISA-95 standards are a generalized model. Because most MES vendors have used the ISA-95 models as a template, there will be considerable similarity in system data structures. If operations staff and IT developers understand the ISA-95 objects models, it will make it easier to understand the differences between MES, and it will make it easier for operations staff to communicate with IT developers when discussing requirements and customization designs.

Machine Learning: An Introduction

To be clear, this is not a lesson on machine learning or AI. These complex subjects are beyond the scope of this book. However, I will provide some examples of how machine learning can be used in a manufacturing plant to help with continuous improvement.

First, recognize that without a well-defined and executed program for process management, getting to the level of analysis outlined in my examples will be a nonstarter. If there are many different incidences of variation, there will be too many variables for a manufacturing engineer to track, and this will result in nothing prominent being determined from a machine-learning perspective. For a manufacturing engineer to be capable of implicating something of significance, the number of incidences of controllable variations must be reduced (look for most processes to be well into Level 2 or Level 3 in the process capability maturity model—PCMM).

Revisiting the material curing oven example, I present some concepts to highlight the manufacturing engineer's line of thinking and an example of how machine learning can be used in analyzing the problem.

The first step is to isolate a problem to be investigated. (Keep in mind that this is a hypothetical example.)

Using a Pareto chart, we can determine that material is not being cured completely, resulting in the product becoming deformed after being removed from the curing ovens. (Although this example is abstract, it provides enough detail to reflect on.)

An Ishikawa analysis reveals a list of the possible causes of the problem:

- Variation in oven temperature control while the product is passing through the oven

- Variation in conveyor speed during the same period

- Variation in the "mass" of a series of products that are passing through the oven

- Variation in the spacing between the products when placed on the conveyor (which may be contributing to variation in total mass)

Depending on the amount of variation that is of concern, there are some other factors that can contribute to the problem, but I will leave the list as it is.

After identifying the list of issues that can contribute to the problem, it is necessary to determine how to measure these variations.

- **Variation in oven temperature** – Use an automated sensor with a historian database to record enough data to monitor the full duration of the product passing through the oven.

- **Variation in conveyor speed** – Use an automated sensor with a historian database to record the actual rpm of the motor driving the conveyor.

- **Variation in product mass** – Take a weight measurement of each product being placed on the conveyor after the material to be cured has been applied. (Because this measurement would probably increase the cycle time a little for this step, this would likely be a temporary measure.)

- **Variation in product spacing** – Visually inspect with a camera to measure the distance between products in real time (*yes, we are getting fancy here*).

Recording this data about the product moving through the oven is a key function performed by the MES. Although programmable logic controller (PLC) systems can record the data, saving the context of which product and when, along with the other parameters to be recorded later in the process (e.g., when a nonconformance occurs), is an important function that PLC/SCADA (supervisory control and data acquisition) systems cannot do easily.

While recording this data, it is important to collect sufficient samples to gain a reasonable characterization. The issue with machine learning (and AI in general) is "the more data, the better" up to a point. As with any statistical study, a larger sample size is better (in manufacturing, a sample size greater than 200 samples is usually good for process characterization), and there comes a point when there is enough data to establish a process characterization and any additional data is found to be redundant. However, the context of "what is being examined" must always be accounted for.

Machine Learning: Clustering

After running production for a while (greater than a minimal sample size), it is time to start looking at the data. The first step in the analysis process is to use a machine learning function called *clustering*, which (in this case) looks at the data and identifies the existing nonconformances, counts the number of occurrences of each type of nonconformance, and determines the occurrences and counts of all other parameters. It is also a good idea to present this output as a Pareto chart to highlight which nonconformances have the highest occurrences. Assume that the nonconformance of "deformed product" has the highest occurrence.

One could get into a discussion at this point of the difference between classification and clustering, but that is outside the scope of this book.

Machine Learning: Nearest Neighbor

The next step is to look only at the records of the product that have the correct nonconformance ("deformed product"), which we will call the primary parameter. Next, we look at variations in all the other parameters (secondary parameters) in relation to the primary parameter that we are measuring and then count the occurrences of each of their values (a variation of the machine learning function called *nearest neighbor*). It might help to present this data in the form of a Pareto chart as well.

The purpose of nearest neighbor is to determine which of the secondary parameters has a (relatively) high correlation to the primary parameter when compared to the others. If a single secondary parameter results in a nonconformance, that parameter will have a higher correlation to the primary than any of the other parameters. Recognize that the secondary parameter is not likely to have a 100% correlation because of smaller influences from other parameters (including some that are not yet being measured). Also, note that having a high *correlation* may not represent *causation*. If no parameter stands out, it will be necessary to run another analysis using combinations of the other parameters (pairing two of the secondary parameters and comparing the correlations of combinations to each other), again, looking for a higher relative count than other combinations. When a secondary (or combination of secondary) parameter(s) stands out, you have only determined an "indicator" of where the problem is. Further analysis is needed to determine the actual cause of the issue.

In an effort to simplify the explanation of nearest neighbor analysis, up to this point, I have ignored some factors in the analysis. Reviewing the occurrences of the secondary parameters likely will require a comparison of the values of the secondary parameters. So, in the example of the curing oven, it will be necessary to look for a *range* of oven temperature drift and not just a drift that is out of specification. This is why a computer (not just a spreadsheet) is needed to perform this kind of analysis. At the same time, it is important to use expert judgment from a manufacturing point of view to guide the data analytics. All this leads back to the concept of having a diverse CoE that works together to solve these kinds of issues.

When reviewing manufacturing data sets, many more data analytic techniques are available, such as training a system's data model via supervised or unsupervised training and many other techniques. As stated earlier, this is not a book on using AI in manufacturing. Which AI technique is best depends on the specific circumstances being investigated. This is also why a data science specialist must be on the CoE team.

Going from Symptom to Cause

In the hypothetical example of the curing oven, the issue was the product becoming deformed because the material was not being cured properly, and we will assume the Nearest Neighbour analysis shows these results from the combination of the oven drifting lower in temperature *and* the conveyor system drifting to a faster rate of speed. Both factors combined caused the material to not cure completely. So now we have solved the problem, right? Well, not yet. For many people, the drifting rates would be considered the root cause and a job well done. But without going deeper to determine *why* these pieces of equipment are drifting in the first place, the team is likely to mitigate the problem rather than fix the problem (find the root cause). Questions that must be answered now include the following:

- What is causing the conveyor to drift faster?

- What is causing the oven to drift lower in temperature?

- Is something happening from a process perspective that is causing the drifts?

- Are the drifts within the equipment vendor's specifications?

The problem with implementing a mitigation is that there is always the likelihood that a different change in the process (for continuous improvement) will interact with the mitigation. Not taking the analysis deep enough is an error that many teams make in continuous improvement, and the actual root cause is not determined. This results in the problem re-presenting itself whenever the controls around mitigation are changed. The other issue with mitigation is that the controls that are put in place to manage the problem are to mitigate a specific scenario. If the issue for the ovens was that the temperature was drifting out of specification, the easy solution would be to tighten the range of control for the oven for that product. However, if the manufacturing company has many plants using the same oven, the other plants are likely subject to the same possible problem. The correct solution may be to have the oven vendor upgrade the oven controls to ensure better performance and then implement that upgrade to all ovens within the company. This eliminates all possible scenarios where this drift may become a problem.

With sensors and equipment (OT) being created to provide more and more operational data and systems like MES (IT) storing more detailed data sets of the processes they control, there is greater interest in the ability to make these technologies communicate more easily (IT/OT convergence, as discussed in Chapter 2).

IT/OT Convergence: Dealing with Data

The key to IT/OT convergence is that large amounts of operational data are now available to IT systems, and the growth of technology and data analytic ecosystems has changed substantially. Distributed data analytic systems such as Hadoop or Google Cloud's BigQuery have made large-scale distributed data systems (sometimes referred to as *data lakes*) available in near real time (within seconds). With increased computing capability and improvements in machine learning applications, these data lakes (large amounts of data in varying formats) can be analyzed by AI systems and provide indicators of different sorts of events at speeds that allow operations management to act before issues become out of hand or before opportunities are lost. As stated earlier, a full discussion on using these systems is beyond the scope of this book and may be beyond the scope of most MOM-supporting systems, but if implementing large scale data analysis, a detailed review of this branch of technology is worthwhile.

Interfacing OT data to IT systems, as discussed in Chapter 2, caused a new problem. The PLC/SCADA systems that monitor sensors in historians can create time-based streams of data that are up to megabytes in size within seconds. From the perspective of understanding the process characterization (recorded by the historian database) of an event back to the IT systems, this stream of data, even in small time frames, is simply too much for the relational databases of IT systems like MES to handle. Many MES have been configured to record an instance of sensor data (taken from a stream of data) as a single datum element, to have some representation of OT events available at the IT level. However, many found that linking an *instance* of OT data to a product quality event during manufacturing was not very effective for monitoring the quality of a process in general. Another error in the decision to try to link an instance of data to a production floor event is that it is not the raw data, in particular, that is important to the events monitored; rather, it is the *relationship of the analysis* on the raw data that is important. As a result, it becomes important to analyze the data and aggregate it in such a manner as to hold on to the important aspects of the relationship to production events.

The reporting of overall equipment effectiveness (OEE) is one of these attempts at aggregation. Note that when using any form of data aggregation, there will be some loss of the important details of a quality event. That is why OEE is used as a monitoring-level data object and not an analysis-level object. However, by monitoring the real-time value of OEE changing, an IT system can recognize the rate and magnitude of the changes and create an alarm to alert to a problem. But to analyze that actual problem, you must go back to the original data stream that was used to create the OEE value. At least with the OEE value being provided in real time, manufacturing and controls engineers have visibility of an issue early enough to go back to the historian for analysis before the event is buried by more data from the OT data stream.

The issue is, how do you aggregate the data? The answer is, as always, it depends!

How to aggregate the data depends on which data you are aggregating and what you are trying to monitor within the process (what the aggregated data is trying to show). It also depends on being able to interpret changes in the aggregated data model. In this section, I look at one such aggregation method and present some of the ways to interpret changes in the data sets.

One of the most common graphic presentations of data from devices such as sensors is the natural distribution curve, or a histogram. In this curve (the number of times a particular occurrence of a value is tabulated), the primary premise is that *if* there is only normal random variation in a process, the distribution of the instances of data will fall into a pattern resembling a natural distribution curve. Figure 4-1 depicts what the distribution of readings from an oven sensor might look like with normal variation when plotted in a histogram.

The key characteristics of the curve from a data aggregation perspective are the mean (average), median (middle value of the data stream), and mode (value with the most occurrences). They are all approximately equal, which can be used as a reference to determine whether the process is experiencing problems. As long as there is an

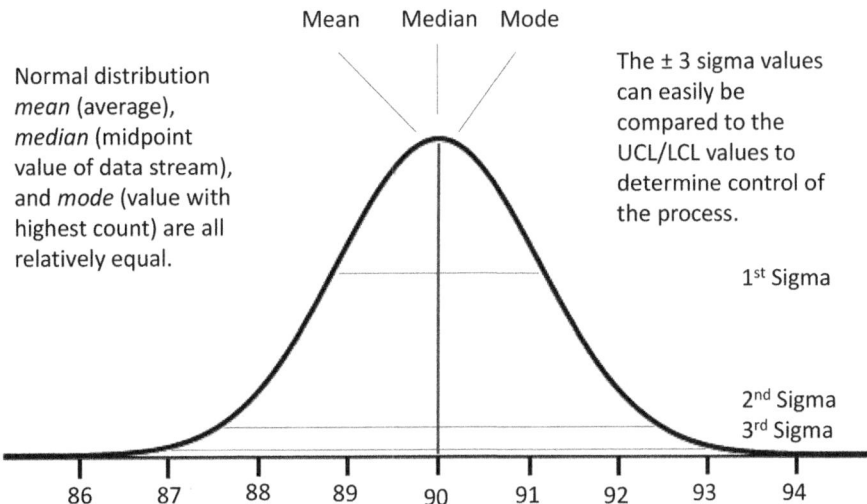

Figure 4-1. Distribution curve of oven temperature showing mean, median, and mode.

indication that the process is in control, there is value in determining the ±3 sigma values and comparing them to the upper and lower control limits (UCL and LCL) and the product specification limits.

An Example Scenario

In monitoring the process of a production unit going through a curing oven, the MES would note the time the unit entered and exited the oven. During this time, the PLC system (a smart sensing device) creates a constant stream of temperature sensor data that is stored in a local historian database in the sensing device. When the production unit exits the oven, the MES makes a request to the PLC for an aggregate of the temperature data during that defined period. Even with the limited computing power available at the device level, the device could take hundreds (or even thousands) of data points and provide a set of aggregated data back to the MES instead of providing a single instance of the sensor data (as is usually provided) or trying to process the entire data stream. (Some systems may use what is called *edge or fog computing* to create the aggregated data set instead of the device itself.)

The following is a data set of aggregated data values for this production unit:

- Mean – 90.0

- Median – approximately 90.0

- Mode – approximately 90.0 as well

- Approximate three sigma distribution – 87.0, 93.0 (to be verified with product specifications and the UCL and LCL)

By doing a quick analysis of the differences between the mean, median, and mode, the MES can look for both acceptable process control monitoring and trending changes from previous production units that have already passed through the oven. If the ±3 sigma values are beyond acceptable limits or the difference between the mean, median, or mode are significant (or trending in an unacceptable manner), the MES will automatically flag the production unit as suspect.

Figure 4-2 shows how these same values would indicate a process that is not in control, although in this case, it is still possible that the ±3 sigma values could be within limits.

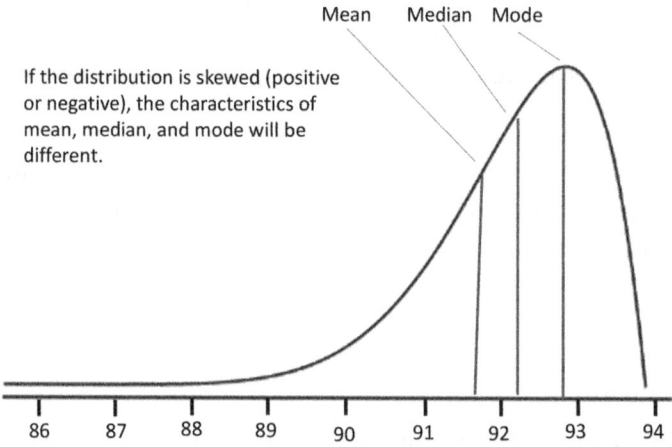

Figure 4-2. Example differences of mean, median, and mode in a skewed process.

Technically, if the distribution is skewed as shown, the ±3 sigma values may not be considered valid because of the skewed nature of the data set. But the immediate analysis of the device data can still be used to determine whether the product specifications have been violated.

The key point is that if this is the kind of data set provided by the smart sensor or if the data is drifting over several production units, it is an indication to manufacturing engineering that the oven is *not* operating with only random variance and, therefore, should be investigated.

Other differences in mean, median, and mode (e.g., mean and median are equal, but the mode is substantially different) can also be indicators of conditions such as a bimodal distribution indicating that multiple process influences are at play or that there has been a sudden change in the process. Again, it is an indicator that manufacturing engineering should investigate the process.

When these data sets are linked directly to the production unit, they can be used with test and inspection results to indicate if there is a link between variation in oven temperature and the failure of the production unit, as discussed earlier in this chapter in the section on machine learning.

Whether the product failed because of the process issues or not, seeing data values drift during a production run can also be an indication of fundamental process management issues that must be investigated.

One of the benefits of MES data is the ability to detect and respond to events in the production environment. An *event* is the occurrence of a special situation that should be given immediate attention. It is important to recognize that the event is a physical action (or reaction) on the production floor. An event could be a sudden breakdown of equipment, a sudden stream of product failures caused by the introduction of faulty material, or a positive situation such as a marked increase in the production rate. When such an event occurs, the cause of the event is not known and must be investigated. The benefit of having these events identified in real time is that it allows people to investigate the issue before the cause is buried in the other activities that are happening on the production floor.

Data Models and Event Management

In the section of this chapter titled "Going from Symptom to Cause," I discussed the difference between fixing an issue and implementing a mitigation. It is always better to fix the root cause than it is to implement a mitigation strategy. However, as discussed in the "Data Sets and Data Models" section, there are times when a mitigation strategy is necessary, and there will be a need to know *when* that event is actually occurring. The purpose of using event notifications (in manufacturing) is to alert operations-level staff to specific occurrences of concern. However, for these alerts to be of value, they must be available (and delivered) in real time. In some cases, if an alert is to be delivered five minutes after the occurrence, that would be too late to act on it. The problem with detecting real-time events is that the system that is using a data model to monitor for an event is frequently also the system managing the production execution. This results in an additional load on the system that is expected to provide real-time response to production activities. In many cases (depending on the event management load), it is best to have a separate system monitoring the data models for events in the production database.

Building an Event Data Model

Assuming that analysis has already identified a problem (or an opportunity), such as the "deformed product" issue mentioned earlier in the chapter, the first step in event management is to understand "what the event looks like from a data perspective." This means one must determine what combination of data accurately describes the event (e.g., drift in temperature, change in conveyor speed, change in the mass being heated). In addition, you must understand what kind of change is significant (i.e., whether the temperature shifts up or down and by how much). After all these factors are determined, using machine learning capabilities, you can then set up and "train" a *data model* to monitor the database for the occurrence of a *data set* that matches the *data model* during production. As part of the growing capability of AI, event management is quickly becoming one of the more prominent uses of machine learning.

Launching the Event

After the data model has been created, the system that is used to monitor the production database (an AI system or an MES module) must monitor the production database while production is running. Most MES applications provide this capability, but if the load of event monitoring is significant, there may be a need for an external system or this function might be provided by many *edge* computing modules that are now included as part of the MES application stack. When the occurrence of a data set that matches the defined data model is detected, the event management aspect kicks in to create an alert for the occurrence, identify the particulars of the data set (serial number of the product, which equipment, and the condition of the current parameters causing the alert), and include all this information in an email (or dashboard) to a response team (part of the CoE). By alerting the response team at the time the data set is created, the team can get to the correct production location quickly to understand the physical conditions that created the event. Depending on the severity of the issue, the alert may be just a notification of the event, or it may include an actual safe shutdown of the line (e.g., stop feeding product into the oven). The actions taken during the event are a matter of management policy.

Many MES manage both a real-time production database and a data warehouse of some sort. (Within the systems that support MOM, both are usually built on an implementation of a relational database.) Because timing is an issue for event alerts, managing events must be part of the real-time production database, whereas it is usually easier to perform the analysis activity for continuous improvement in the data warehouse to make coding the analysis easier.

Summary

The field of data science in manufacturing has progressed considerably in recent years with recognized applications of machine learning, and AI in general, becoming much more common. Applying the technology of the Industrial Internet of Things (IIoT) in manufacturing requires a diverse set of skills and knowledge, resulting in the need for "teams" of skill sets. It is also good management practice to combine these teams under a single management policy. A CoE with centralized skills (data science, program management, quality management) combined with localized expertise (plant-level manufacturing engineering and facilities management) provides a company framework that can lead to considerable improvements company-wide and is significantly worth the effort of management support.

5 | The CoE in Maintaining MOM/MES

In this chapter, I discuss key aspects of the ISA-95 series of standards and how they relate to an MES. In one form or another, there will be a team of people in the manufacturing and information technology (IT) departments that will have a major influence on selecting an MES application. This team will be mandated to become experts on how the MES application will influence the production floor. Therefore, developing this team into a full CoE can significantly help with an MES program in the long term. I also address some key information that a CoE must understand in selecting and maintaining an MES.

Manufacturing Operations Management

Chapter 6 includes a detailed discussion about the functionality of an MES from a manufacturing operations perspective. However, to understand an MES operationally, one must first understand the details of MOM... *(and, no, I'm not talking about your mother, but it is frequently pronounced the same way).*

There are a few industry points about MOM that must be clarified and some misconceptions that must be rectified. First, what does MOM mean? A couple of decades ago, a committee was formed by the International Society of Automation (ISA). It included some major manufacturing companies from around the world and some major enterprise software providers. They started to work on the ISA-95 standards in an effort to help guide the manufacturing industry, software vendors, and consulting firms to create a common understanding of the generic processes that would be needed to manage manufacturing best practices and how manufacturing IT systems were to support those manufacturing management processes (or *activities* in ISA-95 terms). One result was a set of tasks and the relationships between those tasks (ISA-95 Part 3, *Activity Models of Manufacturing Operations Management*). To have a common format to describe these activities, the activity models were presented as a set of *Unified Modeling Language (UML) activity model* diagrams. The highest-level model is the manufacturing management activity model shown in Figure 5-1.

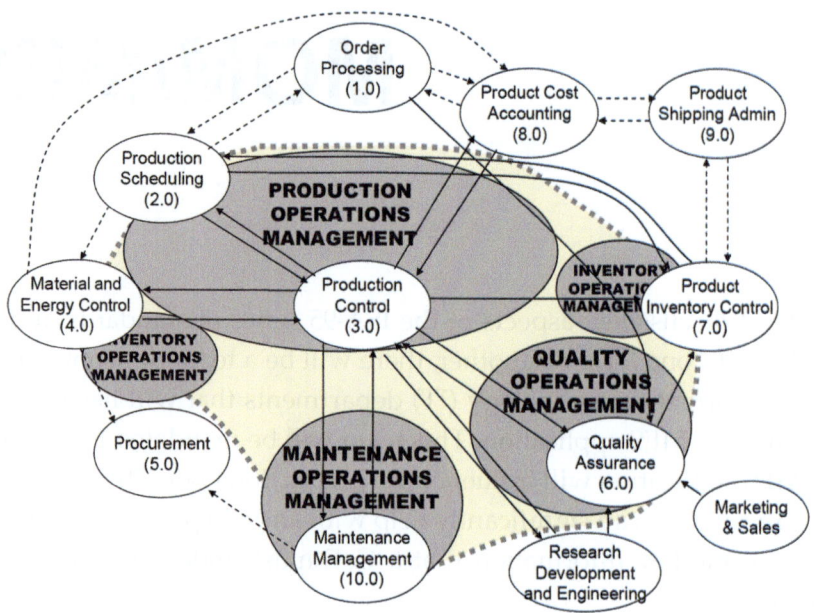

Figure 5-1. ISA-95 manufacturing management model.

Activity Models of ISA-95

This activity model presents 10 groups of tasks (activities) that are needed to manage a company's manufacturing operations. It also includes interactions between research and development as well as marketing and sales, although these functions are not part of the operational perspective. Each numbered oval in the diagram represents a set of specific tasks that are grouped within the context of the company's operations.

> When an activity in the model is not defined by the company, it can be viewed as a gap in the company's defined business processes.
>
> Because it is not recognized in the company with a formal process, it is likely being performed in an uncontrolled and highly variable manner.

In the ISA-95 activity models, relationships are indicated by arrows, that show *dependencies* between these groups of activities. A manufacturing company would use this model for the following purposes:

- Identify the need to specify business processes that may be missing within the context of these activity groups to support the company as a whole

- Define the roles of company personnel who will perform these activities

- Define the possible process-level interactions and communications (not to be confused with IT-level communications between systems) that are needed between groups (meaning the people in these groups must work together and talk to each other)

The model can also be used to help define the requirements that the company's IT systems would need to support. Whether these groupings represent any specific *department* in a company is not determined by the model. That is determined by how a company wants to structure the organization. However, if a company was to compare the model of its *internally defined business processes* to the ISA-95 manufacturing management model, it could indicate the need to define additional processes or add skills that are missing from the company's staffing.

One concern that a company may have is that when an activity in the model is not defined by the company, it can be viewed as a gap in the company's defined business processes. In all likelihood, the company has people who are performing that activity; however, because it is not recognized in the company with a formal process, it is likely being performed in an uncontrolled and highly variable manner.

On the IT side, a software vendor could use the model to plan for functionality that would be needed within the software packages and to define needed IT communication between groups of functionalities within the software.

Another part of the ISA-95 standards is a full set of data objects (ISA-95 Part 2, *Objects and Attributes*, which defines the data models related to processes). This is to be used by manufacturing companies to plan the data collection requirements in a process and by the software vendor to help define the schemas the software package will use in supporting manufacturing. Be aware that a full explanation of the ISA-95 Series of Standards is outside the scope of this book. The full definition of all the parts in the ISA-95 standard series is available at isa.org. ISA also publishes several books that provide guidance on applying the standard.[1]

---

1. *The Road to Integration: A Guide to Applying the ISA-95 Standards in Manufacturing*, second edition, by Bianca Scholten and Dennis Brandl; *Manufacturing Execution Systems: An Operations Management Approach*, second edition, by Tom Seubert and Grant Vokey; *The MOM Chronicles: ISA-95 Best Practices Book 3.0*, Charlie Gifford, editor; and *The Hitchhiker's Guide to Manufacturing Operations Management: ISA-95 Best Practices Book 1.0*, Charlie Gifford, editor.

The ISA95 committee used the manufacturing management model to drill down to lower-level activities. A section of the model was viewed as important specifically to managing manufacturing floor operations and is shown in Figure 5-1 as the light gray (yellow in digital versions). This section consists of the four pillars of MOM (the dark-shaded areas) introduced in Chapter 2: production operations management, maintenance operations management, quality operations management, and inventory operations management. (Inventory is shown as two sections in Figure 5-1 for material inventory coming onto the floor and finished goods inventory leaving the floor). The four pillars of MOM are covered in more detail later in this chapter.

The next activity model to be developed was one containing all the activities that would be related directly to the manufacturing floor (contained within the plant-level management). These activities are directly related to manufacturing or they support the manufacturing activities (i.e., MOM). This activity model is shown in Figure 5-2.

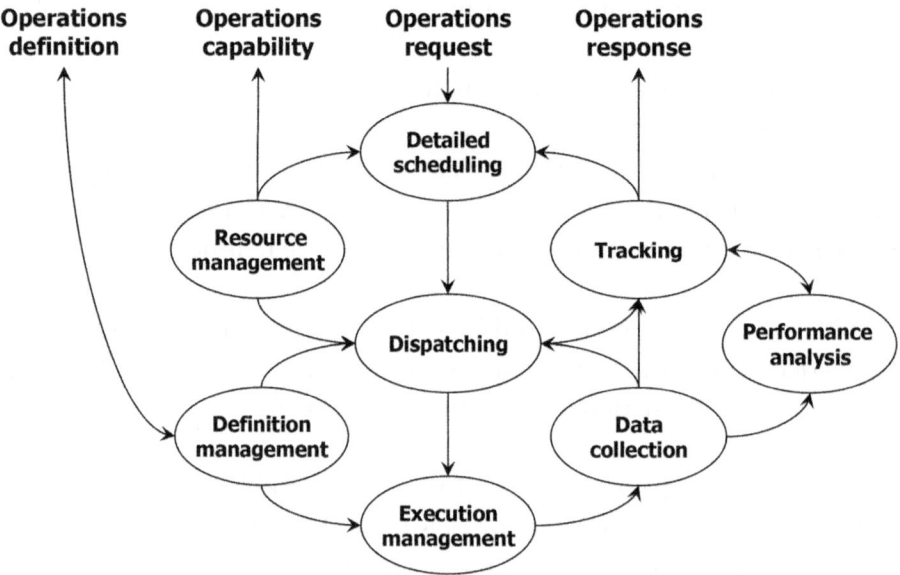

Figure 5-2. ISA-95 MOM activity model.

Working with manufacturing companies globally, the ISA committee developed this generic activity model to cover the activities that are needed to support a company's manufacturing operations. As with the model in Figure 5-1, there are greater drill-downs in the activity models that describe each section of MOM in greater detail. So, what is *correctly* referred to as MOM are the activities, relationships, and data models that are represented in the ISA-95 MOM activity model. That should bring up the question, What is *incorrectly* referred to as MOM? The answer to this question will be addressed later.

Activity Levels of ISA-95

ISA-95 includes the same groups of activities presented in the MOM model but organizes them within levels. By developing this "level" perspective of the activities, the ISA-95 model provides guidance on which activities should be at the planning level (the enterprise), the manufacturing management control level (the plant), and the execution levels (the lines and individual equipment). These levels in the ISA-95 standard are referred to *(with a total lack of creativity BTW!)* as Level 4 down to Level 1, and Level 0 was added to represent the level of actual manufacturing being performed. Figure 5-3 illustrates the activity levels defined by ISA-95 and describes the context of the activities defined at each level. In this diagram, the execution levels (Levels 1 and 2) are separated into two levels and three different models (Level 2 in Figure 5-3), reflecting the different models of manufacturing (batch control, discrete control and continuous control) that relate to my reference within this book to process manufacturing or discrete manufacturing. (Batch manufacturing shares aspects of both process and discrete manufacturing.) Levels 1 and 2 are the process control and supervisory levels that represent the supervisory activities of the production floor and include the physical programmable logic controller (PLC) and supervisory control and data acquisition (SCADA) systems. The activity group referred to as MOM is the primary group of activities within Level 3. Level 3 is also the primary level of support that is provided by an MES. As a result, Level 3 is sometimes also referenced (although not officially) as the MES level.

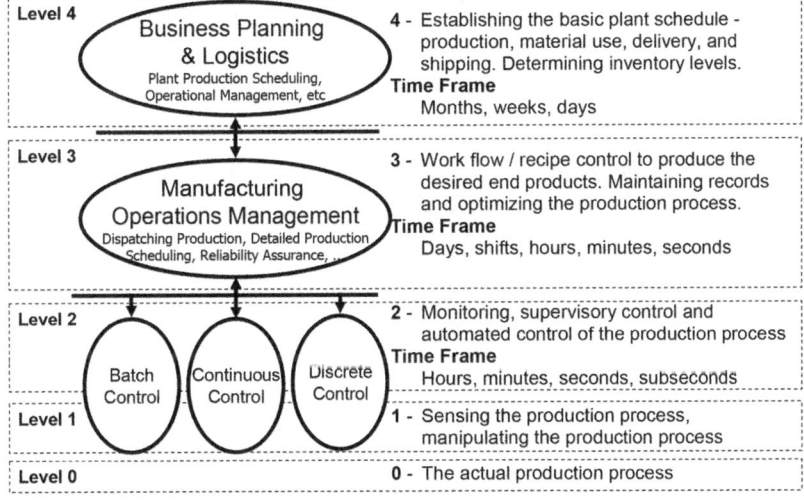

Figure 5-3. ISA-95 functional hierarchy.

Source: Reproduced with permission from the International Society of Automation. ANSI/ISA-95.00.01-2010 (IEC 62264-1 Mod), *Enterprise-Control System Integration – Part 1: Models and Terminology*, figure 3.

The four levels in the ISA-95 standard can also guide enterprise IT departments to properly demarcate which level of the stack of a company's systems should support the activity levels. All the business planning and external logistics are to be handled by systems like enterprise resource planning (ERP). Product definitions and lifecycles are also handled at this level by systems like product lifecycle management (PLM). If company business processes are defined (most should be) by where the activities are within the ISA-95 activity levels, the IT department can then design the support for these processes and the interfaces between their systems according to ISA-95 levels. An easy way of determining the appropriate demarcation is to look at the scope of the activity. If an activity is not specific to any particular plant, it probably belongs at Level 4 in one of the enterprise systems. If, on the other hand, the activity is specific to a particular plant, it is probably a Level 3 activity and support belongs in a system such as an MES or laboratory information management system (LIMS). To take that line of thought even further, if the scope of an activity is specific to a specific piece of equipment at the work-center level in a plant, it is probably to be supported by a Level 2 system like SCADA.

The Four Pillars

In Chapter 2, I introduced the four pillars of MOM (although some have mistakenly referred to them as the *four pillars of MES*). Over the last decade, MES have expanded their capability to support the full function of MOM based on the four pillars. To understand how to support MES functionality, we first must understand (to some degree) which activities an MES supports within MOM. However, several sets of activities are presented as a common model (more or less) that is duplicated in all four of the MOM sections. Within each of the sections, there are also common operational data structures that provide a common understanding of many of the data models within MOM as a whole.

As in any standard where there is commonality, it is to be exploited. Figure 5-4 provides the structure of the common activities for all sections of MOM. Although there are slight deviations in the actual activities in each section, the model generally holds true in identifying groups of activities needed by the four management sections and the dependencies between the activity groups.

In the following sections of this chapter, I cover the four pillars to help readers gain a general understanding of the MOM environment, and in Chapter 6, I will cover the MES functionality that supports the four pillars. As stated earlier, these four pillars are:

- Production Operations Management

- Maintenance Operation Management

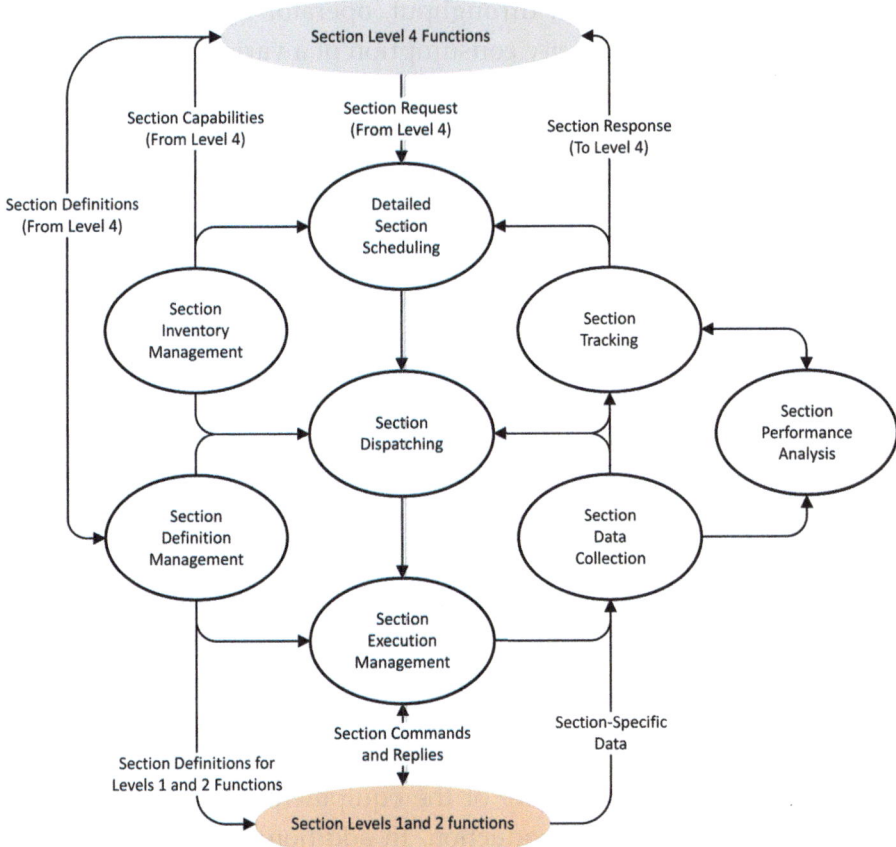

Figure 5-4. General management model for all sections of MOM.

- Quality (Execution) Management

- Inventory Operations Management

Production Operations Management

The ISA-95 model for production operations management identifies the general activities and the data models that must be performed in scheduling, managing, and controlling production orders. In a structure similar to all the management models, the production operations management model is further broken down into subactivity models with the four additional elements of product definition, production capability, production scheduling, and performance management.

In the ISA-95 activity model, production management manages the *available capacity* of a manufacturing environment's resources to deliver on planned customer requirements. In managing available capacity, the planning group must estimate the available capacity (equipment throughput, operator training, and availability), compare it to the estimated capacity consumption of a variety of products, and then develop a plan that uses these resources effectively to deliver customer orders in an expected duration of time. Production operation management also includes the monitoring, reporting, and analysis activities that help improve the utilization of available resources.

Maintenance Operations Management

The ISA-95 model for maintenance operations management identifies the general activities and data models that must be performed to coordinate, direct, and track the functions that maintain all production equipment, tools, and other related assets to ensure they are available as required by manufacturing and to schedule maintenance (reactive, periodic, preventive, and proactive). In a similar structure to all the management models, the maintenance operations management model is further broken down into models of four additional elements: responding to equipment problems, scheduling periodic maintenance, providing "condition-based" maintenance according to data collected on equipment states, and optimizing operating performance and asset efficiency.

This model manages the *total capacity* of the equipment and resources to maintain the planned capacity needed by production. In addition, maintenance management includes managing the planned delivery of equipment *capability* to perform the required manufacturing tasks, which depends on the original equipment specifications and the consistent maintenance of that equipment. A key to maintenance management is balancing the frequency of maintenance programs and the cost of breakdowns from the equipment. Maintenance management includes initiatives such as total productive maintenance (TPM).

Quality (Execution) Management

The ISA-95 model for quality operations management identifies the general activities and data models needed to ensure that resources are qualified and available to track, measure, and report on the quality of the material going into a production environment and the products (or subproducts) that are released from a production environment. This includes all activities for managing product quality, testing and qualifying materials (raw, intermediate, and final), and testing and qualifying all resources (equipment and people). It also includes qualifying material and the dispositioning of material that does not pass testing. In addition, the analysis of material, equipment, and processes as defined for a continuous improvement program and the management of that program are defined as part of the standard.

The concern for quality from a MOM perspective is ensuring that the manufactured products meet customer expectations. As part of quality assurance, quality engineers must also implement inspection and test procedures to ensure products meet design specifications. However, as a result of normal variation in production processes, there must be a balance between the cost of improving the production process to remove normal variation and the level of quality achieved.

An alternative method that more advanced companies have adopted is to measure process performance characteristics and ensure that these characteristics show that processes are "in control" and minimize the possibility of variation (and, therefore, defects). This enables quality execution management to reduce the level of inspection and/or testing to audit levels and, in turn, reduces the cost of manufacturing.

Inventory Operations Management

The ISA-95 model for inventory operations management identifies the general activities and data models that ensure the identification and availability of material entering the production environment and the identification and availability of the end product being produced. This includes all activity for managing the inventory itself and the levels of inventory needed to ensure availability. The ISA-95 model does not specify whether the inventory methodology should be batch or just-in-time or some other methodology, but it provides guidance on the need for a methodology.

Inventory for the supply to production and the finished goods is part of the management requirements for MOM. Managing inventory is ensuring the minimal amount of investment into inventory while ensuring the *expected* availability when needed. Whether inventory is on a shelf in a warehouse or stored as line-side stocking (LSS) at a production workstation, keeping inventory will cost money for storage, and higher inventory increases the risk of obsolescence to the company. Storage cost is not recovered until inventory is sold as finished goods. From a MOM perspective, the concern of using premium space on the manufacturing floor (space that may be used for value-added activity) must be balanced with stopping a production run as a result of LSS inventory not being available when needed.

Summary of MOM

When you understand how these management functions interact and support the full capabilities of MOM, it becomes easier to understand the functionality of an MES in supporting these manufacturing management functions.

The ISA-95 model does not define which IT systems are used in any of the levels.

MOM (the ISA-95 interpretation) is a collection of activities and/or processes and their relationships that provide access to operational data for managing operations. (Operational data is not to be confused with IT data.) *It also includes communicating the operational data between these activities and processes that is needed to manage the full scope of manufacturing operations within a company. In addition, MOM includes a demarcation of responsibility as to which of the four levels of operations is responsible for those activities, processes, and data.*

To be clear, the ISA-95 model does not define which IT systems are used in any of the levels (e.g., Level 4 may be an ERP, PLM, or another enterprise-level system), and there may be multiple systems at Level 3 that support the four sections (pillars) of MOM (e.g., LIMS or MES). In addition, how the IT data represents the required operational data is a matter for the specific implementation. However, each of the section activities must be supported. In addition, each of the activity groups will then drill down into more detailed activities or there may be more activity groups that are specific to each section. For example, the activity group "Section Definition Management," when discussed in a production management context, will reference the activities for defining, controlling, and modifying the *product* data that will be identified in a production order and the definition of the *production order* itself. That same section activity, although very similar, will reference *inventory* identification in an inventory management context. However, both production and inventory management will have a similar modeling structure as well as similar dependencies and data model structure from a manufacturing operations perspective.

As discussed in the previous sections, the ISA-95 activity models are used to define the *requirements* of manufacturing operations processes and the general data structure of each of the four management sections. It then becomes the responsibility of the CoE to define these processes and ensure that the IT systems can support both the defined processes and the availability and reporting of the operations data and information. By using the ISA-95 standards as a template, the CoE is given the necessary guidance to properly define the best practices of manufacturing operations and to define the proper support of these best practices from a management system perspective.

By understanding the relationships between the activities in the ISA-95 standards, initiatives such as ISO-9000 and Lean can be guided to ensure that the quality management systems needed by ISO-9000 and the value stream mapping for process management needed in Lean are defined correctly.

Deciding When to Use an MES

As stated, the ISA-95 standards do not specify any IT systems to be used specifically to support manufacturing operations. As a result, it is possible to completely recognize and use the ISA-95 standards without using an MES. (Remember, the most popular data collection system is still a spreadsheet, a key issue that Industry 4.0 is trying to change.)

However, the issue with using spreadsheets (as production floors quickly find out) is that collecting data in real time is not the same as having access to that data across the entire scope of MOM in real time. And although automated equipment may be able to provide an alert to the manufacturing engineering team when there is a problem, the equipment cannot notify all other groups in the plant and include the context of what the specific problem means to the other groups.

Not using an MES is also an issue when the complete history of a production unit is expected as part of a shippable product. With complex products, such as in aerospace and defense, automotive, and large systems manufacturing, customers are expecting to receive not only a physical product (end product or a subassembly) but also the complete results of all tests and inspections and often the complete as-built record of each production unit as part of the product delivery. This is especially true in the pharmaceutical and medical devices industries. Many products in these industries cannot be accepted by the customer without the full history of the manufacturing of the product. This level of data collection and access to that data will severely overload almost all manual data management programs.

Real-Life Experience

While I was consulting with an automotive company, the controls engineering team was extremely reluctant to allow their automated manufacturing system to depend on the MES to be the main system for managing the production process (*actually they said "it wasn't #@&*% going to happen".*) They started to design a complete automated system to manage production floor activity, whether an MES was available or not. After much discussion (over the course of several weeks), they finally recognized that the controls system could not produce the full scope of reporting needed by their customer to accept the product. The operations management group then estimated that if an MES was not managing the data collection for the process, it would take a little over 100 additional people to collect the data that the MES was handling and the floor would still have a delay of several hours to compile all data collected into reports to be shipped with the product. So, whether the automated system could produce the physical parts or not, the product still could not ship without the associated reports. Because this plant was supplying product in a "just-in-sequence" model (just-in-time with an exact sequence of delivery), it would have no capability to support the customer's delivery requirements. The controls engineering team recognized the importance of the data the MES handles and the speed at which the system could provide access to that data, and they finally agreed to a complete redesign of the controls system using the MES as an integral part of the primary process management system.

There comes a time in a company's management of manufacturing when handling the complexity of the manufacturing process, the data being collected, and the speed with which the data must be available requires an MES. It is needed not only to maintain the ongoing production process but also as part of the analysis processes described in Chapter 3.

In addition to the requirement for real-time data collection and for the speed to make the data available, the complexity of process interactions also increases the need for an MES. In the early stages of a CoE, a lot of effort will be expended to stabilize processes and make them repeatable. This will improve the visibility to the major issues that inhibit a production floor's optimization goals. These issues are usually directly related to specific equipment and the operating parameters of that equipment in relation to the specific production unit being manufactured. There will also likely be an issue with the interaction that different operators have with the equipment they are working on (also related to the stabilization of processes). As the focus of these analytics is highly contained to the specific equipment being investigated, comparing the results of changes to these specific equipment parameters is still relatively simple. However, as the investigations into problems go deeper into process interactions, issues will arise from the interaction *between* different pieces of equipment within the process. It then becomes much more important to have the details of equipment parameters collected and maintained as part of the device history of the production unit that an MES provides so that the interactions between equipment can be investigated.

Considerations for Selecting an MES

Chapter 6 covers material that is directly related to selecting an MES and discusses how the function of an MES will vary depending on the various attributes of the manufacturing company and the structure of its products. These attributes can be used to create a model of the company's operations, and then operations and product models can be employed to determine which MES application is needed to support manufacturing.

Many MES have been created to support specific manufacturing models and specific industries, and the company's selection team must have a common understanding of these manufacturing models before they start looking at actual systems. It will also be important to understand the potential changes that a company is likely to undergo in the first few years after the MES implementation.

One of the primary differentiators of manufacturing models to consider in selecting an MES is whether the product line is structured for process manufacturing or discrete manufacturing. Although there is also a consideration for batch or continuous manufacturing, my experience is that with these models of manufacturing, there is still a fundamental "process" or "discrete" base to the manufacturing model.

Process versus Discrete Manufacturing Models

The primary difference between MES designed for process manufacturing and those designed for discrete manufacturing is the main focus of the database objects and the MES functions that are developed to process these data objects.

Process-based MES are given specific equipment and configurations to use to create an entire batch of "bulk" product, which will then be separated into smaller end-quantity units according to production orders. But the bulk quantity is the primary identified product that is manufactured and tracked (this is the main focus of the system schema), and the runtime parameters of each piece of equipment used are recorded against the bulk product identifier.

Discrete-based MES define operational segments that will track the production of the product segments independently into distinct end products. Therefore, the data for the individual production unit (not a bulk quantity) will be the main focus of the system schema.

More details on process versus discrete manufacturing and the MES are provided in Chapter 6.

Horizontal versus Vertical Manufacturing Alignment

Another aspect of a company's manufacturing model that will affect both the CoE and the selection of an MES is whether a company is vertically aligned or horizontally aligned.

A company that is vertically aligned (as shown in Figure 5-5) has multiple plants (or multiple lines) under the same corporate umbrella, each of which is used to manufacture a portion of the end product. Additional processing on the end product (or subassemblies) at subsequent plants will continue until the final assembly plant completes the end product for sale to customers. These plants are likely a combination of models, with some being *process-based* and others *discrete-based* manufacturing, and it is important for the CoE to maintain focus on the entire production process throughout all plants. Because the company has full control over the end-product quality, the CoE will want to ensure that all processes throughout the plants are optimized for the best outcome of the end product from final assembly. This is

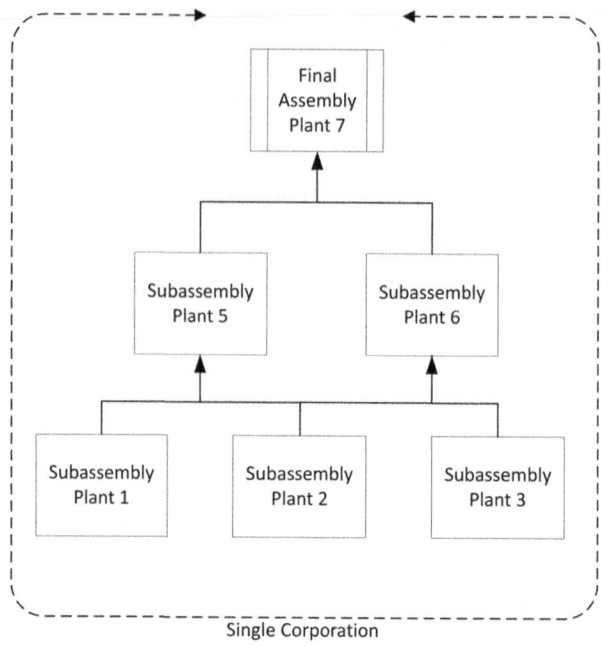

Figure 5-5. Vertical alignment of plants.

similar to a company with multiple lines in a plant, with each product being fully manufactured through an individual line but in one plant.

On the other hand, a company that is horizontally aligned (as shown in Figure 5-6) has multiple plants, but each plant manages the manufacture of a subassembly (or a portion) of the end product for an external original equipment manufacturer (OEM). This is similar to the model used by most automotive parts and subassembly manufacturers that supply OEM automotive companies, and it is the manufacturing model typically provided by contract manufacturing companies.

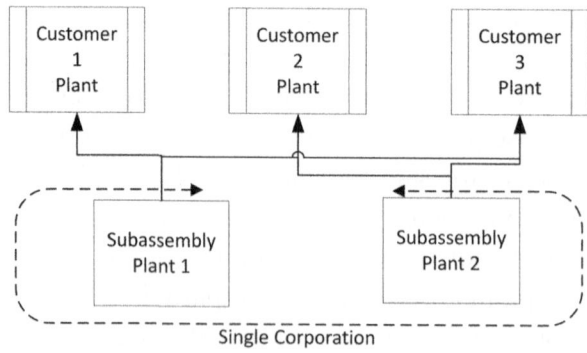

Figure 5-6. Horizontally aligned plants.

In this manufacturing model, the company has limited visibility of the quality of the end product and has ownership of quality only at the subassembly level. In these companies, the primary model of manufacturing is usually discrete- or process-based. Based on a company's defined expertise, it will provide similar manufacturing knowledge for many different end products (e.g., making shock absorbers for many vehicle producers). There will be elements of the CoE that are best to maintain centrally; however, because each plant only owns the quality of the subassembly being manufactured there, the plant is able to focus on decision-making at the plant level instead of at the end-product level. Although it is still important for the CoE to maintain a segment of skills at the centralized level, the need for a full scope of the end-to-end process is not as great. Companies that have many products, each of which is fully manufactured at a single plant (or single line), also fit into this model.

For the structure of the CoE, some differences between horizontal and vertical alignment must be considered. In a vertically aligned company, there is a stronger need to have process and operations knowledge at the central level. This centralization is required to maintain focus on the end-to-end process and on the quality of the end product. The central team also helps to coordinate activities (analysis, solution development, etc.) between plants, and it can be more readily shared by plant-level CoE members that are product-line specific.

In a horizontally aligned company, there is much less focus on the end-to-end process of the end product. Some of the primary leadership and IT resources can (and should) be at the central level. However, the focus from the process perspective should be more on product types and shared manufacturing technology. One example is in the automotive industry, where a company that makes dashboards at different plants and for different vehicles may be able to share the process and technical knowledge and create common support solutions for manufacturing IT requirements.

From the perspective of the MES application, it is important to understand the full scope of required visibility. In a vertically aligned company, there will be requirements to have the data from one plant, such as build records or device history reports (DHRs), available to all other plants. These vertically aligned companies may need IT aspects such as data warehousing to be accessible to the full CoE team across all plants and the corporate office. Many companies try to satisfy all plant requirements with a single MES product. This is frequently a result of the company's IT team driving the selection and implementation process. If the company has implemented a service-oriented application (SOA) system to handle the enterprise integration, having a single MES application can be a severely limiting mistake. By limiting the production environment of plants that use different manufacturing models (and, therefore, have different

MES requirements) to a single MES application, the resulting MES implementation may lack the key functionality needed by a specific plant, or a significant amount of money will be spent on customizations that still only marginally support key requirements. If an SOA has not been implemented, the cost of supporting multiple system interfaces may be cost-prohibitive. However, implementing an SOA system, even at a late stage, may still be the best direction. SOA interfacing is covered in more detail in Chapter 6.

Having free access to the data of different levels of the build process can make solution development much more effective. It can also provide an opportunity for plants to lower the cost of IT services by sharing installations of the MES application. There are configurations for installing an MES application that may permit a lower cost to support and provide catastrophic failover capability. These options depend on factors such as the relative (cyber) distance between plants and differences in manufacturing models.

When a company is horizontally aligned (providing similar but limited subassemblies to diverse customer companies), the perspective of data movement changes considerably. Depending on the contract with the customer, communication of the build history in an advanced ship notice (ASN), for example, may be limited to the serial numbers (or batch IDs) that are included in a shipment or may require a more detailed report. The details and delivery requirements of the ASN can have a significant effect on the interface requirements of the enterprise as a whole. In this business model, the manufacturing and quality engineers have limited access to the data required to determine the root cause of an issue and, therefore, the solution to be implemented. If the company has multiple plants that produce similar products, it would be beneficial for the plants to share insights and information, but only if they have similar manufacturing models and MES configurations supporting those models.

As the actionable scope of the CoE will likely reside at the plant levels, the CoE will also have to ensure that the training for the full scope of the MES functionality (even functionality that is specific to a single plant) is also recognized by the CoE team members at all plants.

Developing the Requirements

One of the first areas where companies make mistakes in selecting an MES is in creating the requirements to determine which application is best. If the company is entering the MES selection initiative for the first time, there will likely be a lack of understanding of MES functionality and what the priorities for selection should be. There will also be a lack of understanding of how the manufacturing floor will change (resulting from the initial implementation of an MES and from changes in manufacturing in general because of improvement initiatives over time). Common mistakes I have encountered include (1) having an IT-centric team lead the requirements-gathering process (in most cases, the team does not understand the full scope of manufacturing and fails to ask the correct questions of their manufacturing counterparts) or (2) creating a set of requirements to favor a specific MES application because someone within the team has a relationship with a specific MES vendor or an implementation consultant specializing in a certain MES application, or because of the perceived cost. In either case, the result is a set of requirements that do not reflect the full scope of manufacturing needs, and the MES application will be selected based on requirements that are skewed and do not properly reflect the needs of manufacturing operations.

Therefore, it is important to have an established CoE before developing requirements. Many companies will have already implemented Lean initiatives or ISO-9000, and as a result, will already have a team with a sound understanding of manufacturing in general. The team that has played a leading role in one of these initiatives (assuming the teams for these initiatives already include process management and quality management specialists) would then be a good starting point for an MES selection team (and a start for the MOM-level CoE as well, if it is not already available). This team will still need to be augmented to include some IT knowledge (and data mining knowledge), and it would be advisable to include some knowledge of controls engineering as interfacing with equipment on the production floor will be required at some time.

It is important for the MES implementation team to establish a clear scope and purpose for implementing the system. A frequent mistake in MES implementation is to reference a single plant for the requirements. Often, the steering committee for an implementation selects a specific plant to be the "test plant," and the team will only look at that plant to determine the list of requirements. However, this narrow scope is usually a result of an assumption that each subsequent plant will independently determine its own scope for the implementation, despite a primary requirement to use the MES application that was originally selected for the first plant. The "subsequent plant" assumption and referencing a single plant for requirements will result in costly implementation projects for each additional plant. If a company has already established a common CoE across all plants, this mistake in the initiative scope can be avoided.

A normal *(or it should be normal)* part of project management is to establish the goal of the MES initiative. The goal should roll up into one (or a few) of the corporate strategies to ensure that there is an understanding with senior management as to what to expect from the MES initiative.

If a company already has a CoE for manufacturing, an additional input into the requirements should be to define how the MES system will help the CoE. Whether the CoE was established as part of a previous initiative (e.g., Lean, Six Sigma, or ISO-9000) or is new for the MES implementation, the team should understand how the CoE will use MES in the long term.

If a company is new to an MES, determining *what* should be covered in the requirements can be confusing. The ISA-95 activity and attribute models can be of great help in this regard. By using the ISA-95 activity model, a CoE can map out management processes that may not have been defined. If the activity model references an activity that has not been defined in the requirements, the CoE has an indicator that something is missing from the requirements. As stated earlier, the missing requirements should be interpreted as missing processes the company must define. Whether the process is defined before or after the MES implementation, ensuring that the system is functional for all requirements (even if some system functions do not fully support a requirement) will help to reduce the cost of implementing customizations at a later stage.

By using the ISA-95 activity and attributes model as a guide, a company will know which operational data should be reflected in the MES and how different data objects should relate to each other. This will provide an indicator as to how well an MES can represent a virtual production floor and how easily the reporting requirements can be supported.

> As a company gains knowledge in MOM and MES, the processes and requirements will change. Being aware of this fact, and accounting for it up front, will save a lot of headaches in the future.

In addition to identifying process and functional requirements, the ISA standards can be used to identify training requirements for operational staff to understand the purpose of the processes and how they contribute to manufacturing operations and for training in MES functionality.

I have stated this before and will state it again throughout this book: As a company gains knowledge in MOM and MES, the processes and requirements will change. Being aware of this fact, and accounting for it up front, will save a lot of headaches in the future.

Selecting and Implementing the System

At the time this book was written, there were well over 300 different commercial MES applications on the market. These are all collectively referenced as *commercial-off-the-shelf*

(COTS) applications. They include a mix of design concepts that support either a wide range of manufacturing models, a specific manufacturing model, or a specific industry. These applications do not include the hundreds of attempts by individual companies to design their own MES applications internally, which are referred to as *homegrown* systems. In this section, I outline some of the key aspects of each type of system design and look at the pros and cons of each.

Understanding COTS versus Homegrown

Over the last couple of decades, many global manufacturing companies and organizations (e.g., ISA and MESA International) have pooled their knowledge to develop and implement best practices for manufacturing operations. In the process they have been gaining a considerable understanding of the processes and data requirements to support manufacturing operations. As mentioned earlier, it is from this collaboration that the ISA-95 series of standards (including the ISA-95 MOM activity model presented at the beginning of this chapter) was developed. MES applications have been on the market since the late 1990s. During that time, the knowledge base for developing COTS MES applications has grown substantially, and vendors have had the opportunity to evaluate many design concepts and expand on the delivered functionality that supports a broad range of activities in the ISA-95 MOM activity model presented in Figure 5-2.

As a result, the COTS applications are becoming fairly well designed and quite capable in performance. COTS application developers have also moved away from needing continued coding to incorporate required functionality by making these systems highly configurable to provide small changes in applying business logic rules. Although this level of configuration can be both good and bad, when managed by a well-trained staff, these systems have the potential to become very useful and robust. An important consideration while implementing these COTS applications is that they have been designed to support industry-recognized best practices, with some aspect of configurable deviation in the best practices. As long as a company has implemented processes based on these best practices, most of these COTS MES applications will be able to perform as required. In well-established industries like consumer electronics and appliances, these best practices are relatively easy to adopt and need little augmentation. On the other hand, in industries that are relatively new (e.g., some green technology companies), the industry best practices have not been fully vetted and are, therefore, in a state of flux. Designing MES applications to support some of these industries is a bit more of a challenge. This, of course, is one of the reasons for a well-established CoE.

An outcome of the industry collaborations mentioned earlier is that there has been a considerable amount of cross-industry knowledge transfer, and best practices in one industry are being found to be adaptable to other industries. Following a similar spread of knowledge within a company's CoE can only help to make the CoE more effective.

An additional characteristic of COTS systems is the support of best practices in software development lifecycle management (SDLM). The majority of application vendors will adhere to these best practices and maintain consistent *look-feel* policies and coding standards, as well as design initiatives that are well documented and well managed at a project and design level. As a result, with COTS applications, there is now a considerable amount of consistency in the functionality delivered by COTS vendors that correlates with the ISA-95 standards.

A significant problem with COTS applications is that these design initiatives and supporting industry SDLM best practices also come at a cost that is reflected in the application licenses. Another potential problem with COTS applica-

> If the application is designed to support best practices in a company's industry, is there really a need to customize?

tions is in the lack of ability to support nonstandard processes within an industry. Most applications provide a means to support customizations (and there will be customizations), but depending on the vendor and the nature of the implementation (more on this soon), the ability to support a wide range of customizations may be limited. Questions that should be reflected upon when evaluating the need for customization are: *If the application is designed to support best practices in a company's industry, is there really a need to customize it? (That is, should the company adhere to best practices, or if best practices are not suitable and customizations are required, to what extent should the application be customized?* These are real concerns that a CoE must consider when the requirements do not quite align with the application's functionality. Note that depending on the nature and number of the customizations, it can be very costly to upgrade a current version of an MES to the next revision.

Real-Life Experience

I was working as an MES consultant to implement a new system for a client that was already accustomed to using a homegrown MES application. While discussing the requirements of the new implementation using a COTS system, the client's lead internal adviser made it clear that the new system needed to work very similarly to the older homegrown application from a production user perspective, despite having a very different functional capability and database structure and despite a strong recommendation by our team against this direction. By the go-live, the implemented system functionality was using more than 60% customized function, including a complete overhaul of the production user interfaces at a significantly higher than normal project cost (approximately two and a half times the normal cost). It should be noted that the system was quickly adopted by the production floor, resulting from the very similar functionality. However, because of the cost that would be incurred to refit all the customizations, the system was never able to be upgraded and stayed at the same revision level for nearly a decade. This also increased the cost of support for the system. Eventually, the vendor could no longer support that version, and the client had to go through the planning and cost of an entirely new implementation. The company finally gave up on a majority of the original customizations to try to lower the implementation costs. Although it would likely have required a bit more training on the original implementation, had the company taken a different route, it could have saved tens of thousands of dollars in the original implementation and provided a platform for continued upgrades at a much lower cost of support. Only the company itself can determine if the correct action was taken.

The primary reason companies have provided for developing a homegrown MES has been the cost or the method of funding needed for COTS applications. It is usually far easier to get approval for an internal project that is charged to a company's operational expenses than it is to get approval for a large capital expenditure. These companies did not doubt that some aspect of an MES was needed, but the expense of implementing the larger systems can be out of reach for many small to medium-sized companies. However, there are issues that company directors should be aware of. In most cases, the staff that will be selected to create a homegrown system will do so as *part* of their other full-time activities. This will result in a longer development process, and the person doing the development will be constantly distracted from the design and development activities, which will probably result in some quality problems with the coding. In addition, the developer is likely not to be aware of many MES best practices and will model the functionality on the current practices of the production floor. These issues will cause slow implementation times and will result in what is referred to as *spaghetti code* in reference to a lack of solid structure in the system coding. These issues will impede the company's growth in continuous improvement and process management.

Real-Life Experience

A Tier 1 automotive supplier had several plants in North America and Europe. One of the plants was looking to implement some MES capability and decided to develop a homegrown system. The system was developed and implemented in a bit more than a year and a half and supported most of the plant's data collection and process management needs. The plant manager then approached the head office with a proposal to expand the functionality and implement the system in all plants globally. On paper, it sounded like a good idea. The first indication of problems appeared during the first implementation at a different plant. There were very few common processes at the management or the production floor level; and despite having similar product lines, the second plant had different key performance indicators (KPIs) that needed access to data the system did not currently collect. This required additional functionality to be developed and expansion of the schema for the operational database. The design of the original system did not take into consideration expanding on these issues, and it took more than a year to implement the additional capability. After the company started implementing the system in the third plant, the lack of coding quality resulted in significant coding bugs. The company eventually recognized that expanding the system into many plants would require starting from scratch and recoding the home-grown application. This took an additional three years to complete. Because many of the other plants were eager to implement an MES, they were not willing to wait for the redesign and engaged in either their own homegrown applications or implemented COTS applications. By the time the new version of the original homegrown was ready, many of the other plants already had a system in place and were not willing to switch, making the additional investment into the redesign of the original homegrown system of little value. In the end, the company had a mix of many different MES, no common-ality of processes, and very little gain in long-term value from the implementations.

The preceding example is very common when homegrown MES systems are imple-mented. To make a homegrown system work, it must be designed with a knowledge of the differences that exist between plants, the differences between plants must be minimized, and the coding must adhere to recognized coding standards in SDLM. If these requirements are not recognized at the start, the long-term capability of the sys-tem to support the company will be limited or the system will be an outright failure. Having a well-developed CoE at the initial stage will help considerably in the success of the initiative. Because most of the COTS MES applications were designed and devel-oped with these primary requirements already satisfied, one has to wonder about the value of engaging in a homegrown initiative in the first place. There is some support for using homegrown as an initial proof-of-concept, but when a fully supported MES is required, the value of homegrown systems quickly fades.

In recent years, a few *low-code* or *no-code* platforms have tried to gain a foothold in the MES industry. For homegrown applications, there may be some benefits to being able to easily develop independent systems. On the other hand, there are many issues that you might encounter. In a no-code environment, a visual development workbench is provided with drag-and-drop implementation. However, the term *no-code* is a bit mis-leading. Although the system platform does not require the developer to actually do coding, the workbench uses predeveloped code structures to develop the system in the background. So, although the developer is not required to be "coding," code is being generated by the platform. One of the primary development requirements of an MES is the need for fast, efficient code. Unlike many other IT system applications, having inefficient coding can make an MES unusable. With the background code development in these "no-code" workstations, the developer is depending on what may be question-able prearranged code structures to create fast and efficient production workstations. An additional concern with no-code platforms is that each production workstation is an independent system. Developing a multiuser system that enables data sharing between workstations and using a central database is much more complicated.

With many of the up-to-date COTS MES, the applications are designed to enable workstation and master data setup via configuration modules that can configure up to 90% of the system without coding. The proprietary, highly efficient code structures will likely make the need for no-code platforms unnecessary.

There are also alternatives to full COTS applications. Some system integration consulting firms that have experience with MES have utilized platforms for SCADA or SOA development to create an MES-like system. These platforms provide fundamental MES capability at a much lower cost to implement. However, it is not always easy to expand the capability of these systems as a company's requirements grow and expansions are usually available only if you work directly with the same consulting firm. Depending on the quality of consulting and the integrity of the relationship with the firm, this may not be a problem.

In addition to deciding whether to choose a COTS or a homegrown system (or an MES-like system created by some consulting companies), other factors must be considered when selecting an MES application, such as whether the company has a team of developers already in-house and whether their skills can be extended to the MES in general. Some MES applications have been designed to be highly configurable (require little coding), whereas other MES provide a "development toolbox" and most of the functionality must be developed (lots of coding) within the toolbox.

If a company does not have an internal development team, a highly configurable system may be the better option. However, the company will then be dependent on the insight of the MES vendor and whether the vendor has a sufficient understanding of the company's industry to provide the correct configuration points. Some systems provide so many configuration options that 80% to 90% of the required functionality is available via configuration, but the application could have five or six different configuration formats to perform a similar function, and each option in the configuration would support a different optimized capability. Understanding the configuration options also requires a good grasp of the operational impact of these options right down to the production floor level, giving rise to opportunity for errors in configuration.

To complicate this matter a bit more, I reiterate the point that a company's requirements at the start of using an MES will likely change as its maturity in process management changes. This will result in a change in configuration requirements over time.

An additional factor to consider for part of this decision relates to the concern I expressed earlier about aligning the selected application to a company's production model. If a company has many plants that build a portion of the end product up to final assembly (or the company is a contract manufacturer that just has different plant-level requirements), a broad range of functionality will be needed instead of a highly focused application that supports a single industry. It will be easier to extend a general-use MES to support specific needs than to extend a highly focused MES to support many needs. If, on the other hand, a company has many plants that all support one primary industry (e.g., injection molding), a much more focused MES would be the better option.

Using Cloud versus In-Plant

One aspect of a COTS MES that must be investigated is how the application will be hosted. Options include hosting on the company's internal server and network infrastructure, using a privately hosted application with a "platform as a service," or hosting via the cloud using either a cloud application provider or a hybrid that combines the cloud with fog or edge technology.

To determine the hosting strategy, it is essential to review the characteristics of the enterprise and the system response time required by the production floor. For example, if a production line has a takt time of less than 1 minute, a system that takes up to 3 to 5 seconds to respond to each transaction can quickly use up as much as 30% of the available cycle time to complete a single production unit at a single operation. This kind of delay would be a complete disaster for production planning.

Internal Hosting

An internal hosting strategy is illustrated in Figure 5-7. After the server and database sizing are established, that sizing is dedicated to the MES application and fixed to the set parameters (even when using a virtual machine configuration). This frequently provides the shortest cyber distance between the application client and the server. Combine the short cyber distance with a well-designed network to support manufacturing, and this, in turn, provides the least amount of delay in processing communication requirements. Although many other factors can have an impact on transaction processing speed, such as the efficiency of the coding structure (the algorithms) and the database schema design, these issues will not add to fluctuations in performance as much as a poorly implemented network will.

Depending on the markets in which the company sells product, there may be wide variations in computing resource use with seasonal shifts in production. As a result, in

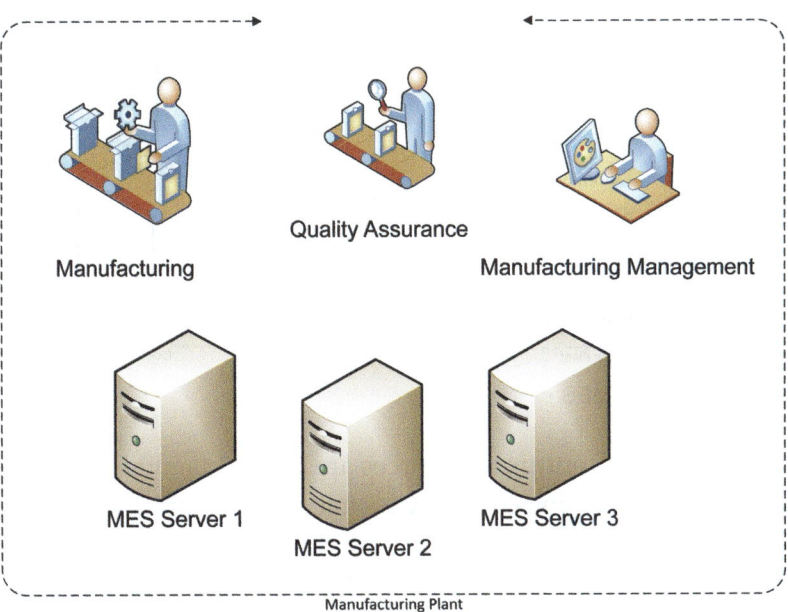

Figure 5-7. Internally hosting the MES.

a privately hosted environment, computing resources may be underutilized (costing more than needed) or overburdened (and therefore slower) as demands on production change for whatever reason. If a company can manage relatively consistent rates of production and growth, the amount of computing resources can be monitored and adjusted as required. In a well-managed company, when there are issues with computing resource availability, it is simple to acquire more computing power or hard drives for the database. With this control, the response time of the MES application is maintained at acceptable levels for speed. In production environments that are highly automated, real-time response is *critically* important. In more manual production environments, there is *some* leniency with system response.

In addition to response time, another concern over internal hosting is support for catastrophic failure. The potential issue with internal hosting is that the company is responsible for its own protection against full system failure. One of the events that can cause this is a power blackout. This concern must also be reconciled with the protection the production floor has. Unless a company has backup power for the entire production floor, the production area will be shut down with a power failure. Losing access to the production control system as well is of little additional concern.

Cloud Hosting

In companies that have wide swings in production rates resulting from seasonality changes, frequent introduction to new markets, or other reasons for rapid change in production volume, cloud computing may be a better solution (see Figure 5-8).

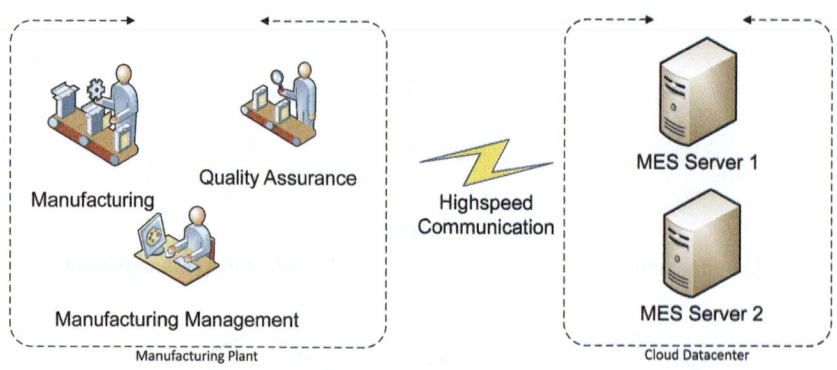

Figure 5-8. Cloud hosting the MES.

In this environment, the computing resources are adjusted automatically by the cloud provider, and the manufacturing company is charged a variable rate depending on the resources used. With the variable "as used" rate, computing resources can be covered under operational expenditure requirements. Under an operational expense, the cost of computing is spread over the life of the MES application. A variation of cloud implementation is software-as-a-service (SaaS) systems, where the computing services are managed externally (limiting the company's staffing requirements), and the company is supported by a dedicated external system.

An issue with cloud environments is that the response time of the application can vary greatly because of forces external to the company. There is also the possibility that the cloud computing center could go offline (e.g., due to a power outage) while the manufacturing plant is still operating. This would, of course, leave the manufacturing plant crippled. However, with good cloud providers, their hot backup systems (used in conjunction with good wide area network—WAN design concepts) would probably be more than capable of maintaining service from an alternate location and likely make these concerns a nonissue.

In an effort to alleviate the performance issue, many cloud providers also provide add-on technology in the form of edge or fog systems (as shown in Figure 5-9). Using these technology options, the cloud system is supported by an on-premises edge or a fog system that handles most of the high-speed transactions of data collection and aggregation (as described in Chapter 4 and also used in event management and KPI dashboarding, for example), which reduces the amount of data (information) going to or coming from the cloud.

As mentioned earlier, important aspects of an MES are the ability to handle customizations, the extent to which a system must be customized, and how the customizations are implemented. Although the extent of customization should generally be minimized,

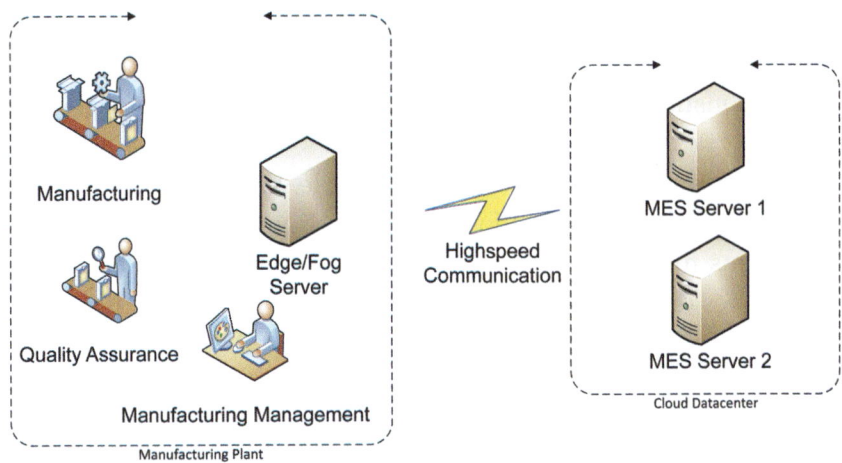

Figure 5-9. Cloud hosting with edge/fog.

plants with unique staffing policies, nonstandardized equipment throughout an enterprise, and different product profiles will inevitably require some customizations. If a plant is implementing a privately hosted system, the ability to customize will be completely in the control of the CoE that is supporting the system. If, on the other hand, the plant is implementing a cloud system, the amount of customization will be limited (if allowed at all). When a change to the application is accepted by the cloud application provider, the functional design from a change request (normally a customization) will be planned into its road map but will take much longer to implement as the design will be included in the system's core capabilities, not as an option only for the requesting company.

> Although there is a strong need for IT knowledge as part of the MES selection process, almost every decision will have an impact on manufacturing operations. Understanding the nature of that impact (from both an IT and a manufacturing operations perspective) will be critical to selecting the correct system.

While investigating which MES is right for a company, many factors must be considered, and the selection of the right application (and the right implementation method) will depend on the insight of the team that is making the selection. Although there is a strong need for IT knowledge as part of the MES selection process, almost every decision will have an impact on manufacturing operations. Understanding the nature of that impact (from both an IT and a manufacturing operations perspective) will be critical to selecting the correct system.

Managing the Implementation

Assume you have reviewed several different MES options (depending on what is appropriate for your company) and have selected the application that is right for the company. Now all you have to do is contract with a consulting firm and have them install and set up the system. (*If only it was that easy.*) Never mind deciding *which* consulting firm; is a consulting firm the correct way to manage the implementation?

What about the implementation methodology (Agile, Waterfall, Lean)? What skill set is required? Should the company allow customizations? What about the interfaces? What other systems must be connected? Which equipment must be connected? The list of questions goes on and on.

Unfortunately, the only correct answer at this point is "it depends." Although some questions (e.g., Agile, Lean, or Waterfall, *although I have yet to see a good Agile MES implementation*) are a matter of preference, other questions should be answered before starting the implementation.

This raises the next question, If the general answer is "it depends," what does it depend on?

In the next few sections, I explain the considerations for many of these questions.

Methods of Implementation

One of the first decisions commonly considered is which systems integration (SI) consulting firm to use. But before tackling that decision, a company should first consider if a SI firm is needed at all. If the company has multiple plants to implement an MES, there are sufficient reasons for that company to consider developing an implementation team internally. The first is that an implementation team of some sort will be needed for a few years (depending on the number of plants, an estimate is about a year per implementation). Second, if the company is using its MES environment wisely, there will be a need for ongoing support and updates that will be required long after the implementations are complete. Any implementation team must develop the skills for the MES or develop an understanding of the company's manufacturing environment, or both. A consulting firm has the skills for a particular MES (sometimes multiple MES). The firm knows the system already and is likely to have some experience in configuring particular requirements, which may or may not coincide with the company's requirements. Then the firm must develop an understanding of the manufacturing environment through requirements gathering. However, requirements gathering does not gain an in-depth knowledge of the company's manufacturing environment. The firm is counting on a somewhat superficial knowledge of industry best practices and the ability of the company's Manufacturing Engineering department to fill in the gaps in knowledge.

If a company has already established a CoE (via a previous initiative such as Lean or ISO-9000), it is an easier task to develop an understanding of the MES application and expand the CoE to handle an MES implementation as well. If there is no pre-existing CoE, it may be of value to use an SI firm for the first implementation. The SI firm can help develop the internal team. Recognizing that the company will need to support the system long term, having the internal team gain a detailed understanding of the decisions made during implementation will help the team provide support. The need to retain the knowledge of implementation decisions can be a key point to consider when deciding who is going to implement the system. When there is no prior initiative from which to draw experience, having an SI firm implement the system is probably the safest option to take. But the company should take a close look at what is being included in the implementation plan established by the SI.

Another decision that must be made early in the initial implementation project is to determine how much of the system to implement as part of the initial implementation plan.

First, some background information is necessary.

COTS MES applications are designed to support the needs of manufacturing companies ranging from those that are fairly new to MES to those well advanced in MES utilization. Depending on the vendor, the application may also support a broad range of industry requirements. However, most companies that already have an MES implemented the basic track-and-trace capability, incorporated some key custom data collection requirements (to support current KPIs), and used the system to help generally refine many of the *production floor* processes (in turn, helping to make these processes more stable and repeatable). After this aspect of the implementation was completed, few companies engaged in implementing further capabilities, leaving as much as 60% of MES functionality unused. This frequently resulted from the initial decision to have an SI firm do the implementation, which also resulted in a slow uptake in MES knowledge internally. In most implementations, the SI firm provided training to internal staff to support functionality that is being turned on but not necessarily the full function of the application, thus causing gaps in knowledge. When looking at the initial scenario of the SI firm configuring a large amount of the application, unless the company already has MES knowledge, much of the configured (and paid for) functionality will still go unused for some time as there can be a long learning curve for the MES and some of the functionality will be forgotten.

In other cases, the SI firm will approach the company's team to initiate interest in implementing a particular function of the system as a result of the latest buzzwords (or the function is a new feature of the application, and the firm wants to see it working) despite the company's lack of maturity in process management. A company that lacks such maturity likely will not fully understand the new feature, and attempting to use the new feature will cause so much frustration that the company will turn the feature off. An SI promoting a feature in this manner leads to investing in an MES configuration and solution design that the company is years away from needing (and may never actually use). It also adds considerable cost to the implementation consulting fees, resulting in a waste of money for the company. This type of consulting has also led to a bad rap for many MES implementations in the industry failing to live up to the hype that the companies were given during the sales process. When a company is already knowledgeable in MES *and* has considerable maturity in process management, making as much of the MES functionality available upfront provides the tools the company needs to move quickly to the next level in improving its efficiency. But this should be balanced with a realistic understanding of the company's growth in process maturity in general; think about the company's process capability maturity model (PCMM) status.

For companies that are new to the MES, it is strongly recommended that the CoE is trained on MES capabilities in general before the application is selected. (MESA International provides good general MES training.) This knowledge will be useful in establishing the requirements in the first place but will also enable the planned implementation of MES functionality that aligns with the company's strategic growth. When the CoE knows the plant's processes and manufacturing operation's process maturity, and they have a broad understanding of the MES capability, the CoE can plan to introduce MES capability as needed instead of configuring functionality that may not be used until months or even years later. It is also strongly recommended (although not necessarily by the SI firm) to implement MES functionality in phases. When an MES is implemented in phases, the CoE (and the rest of the plant) can build knowledge and expertise in some MES capabilities and reflect on the value of other MES capabilities before engaging in the next phase of implementation. During the initial phase, the company can refine many of its manufacturing and management procedures and then better plan for how the next set of MES features will be implemented.

Phased Implementation

Within the functionality of an MES, there is a logical progression of system capability and knowledge development that lends itself to a phased approach that will help with the long-term success of the CoE and the MES program. These logical phases (shown in Figure 5-10) are as follows:

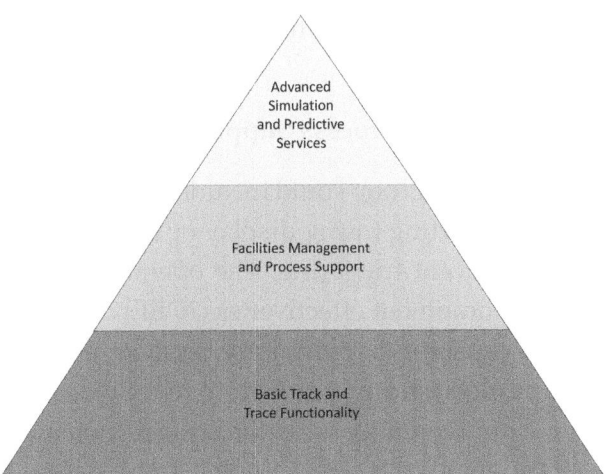

Figure 5-10. Three phases of MES functional support.

- **Basic Track and Trace**

   The primary MES function, track and trace (covered in detail in Chapter 6), is the function of configuring the processes for production units to follow and the resources to perform the operations within the process. The system will then guide the production operators through the operation activities and manage feeding production units through the remaining defined processes. At the same time, the system ensures that all activities performed on a production unit are the correct steps in the process based on its current real-time status. As part of this functionality, the system also records all activity that was performed on the production unit and who performed it. After a production unit has been completed, the system will have a complete history of all completed steps in the process and which resources were used to complete each activity. The system will also have verification that all required activities in the process have been completed.

   During this phase, the CoE is becoming familiar with the MES functionality and the business logic rules designed into each system, as well as the data structures and how these structures reflect the activity of the production floor. If the company is also using the ISA-95 standards as a template for the administrative- and management-level processes, the CoE will gain a better understanding of the relationship between the ISA-95 standard and the interpretation of the system data.

   In this phase, the manufacturing engineering staff can work with the production floor staff and the CoE to understand how processes are defined and managed within the system (ensuring proper representation virtually within the

MES). It is easier to test different configurations to provide optimal support to production during this phase.

- **Facilities Management and Process Support**

  MES functions within this group could include real-time equipment status monitoring, operational reporting (using display options like dashboards or Andon screens), and process event triggering. The operational reporting may include options like overall equipment effectiveness (OEE) or statistical process control (SPC). The primary capability is to monitor, analyze, and report on the status of all aspects of the production environment in real time and provide alerts when anomalies occur. This capability is better utilized when production processes are relatively stable, making it easier to detect anomalies, and requires a considerable amount of interfacing to production equipment. Updating equipment-level communications to Industrial Internet of Things (IIoT) capability will help, but this is not essential. More information on interfacing is in Chapter 6.

  In this phase, considerable effort is required to characterize the processes. This is where the manufacturing and quality engineers gain a clearer understanding of the process and product capabilities, as well as of the relationship between equipment operating parameters and process or product anomalies. Having (more or less) completed the initial track-and-trace phase, there is greater confidence in the numbers the system is providing, and the staff has a much clearer understanding of the anomalies that happen. Provided that there is a structured plan for interfacing with the equipment, these interfaces can be implemented as needed (not all at once). This approach to interfacing with production floor equipment provides the ability to select the critical detailed reports of equipment performance and a platform for initiatives such as OEE and SPC. Interfacing with equipment also enables continuous improvement and total productive maintenance (TPM), all driven by the data derived directly from the equipment.

- **Advanced Simulation and Predictive Services**

  The primary capability in this level of MES functionality is collecting and analyzing the parameters of equipment operations and the product quality during a production run to recognize trends in operational parameters that could lead to equipment failure or product nonconformance (some would say these are the same thing). This function would then alert operations staff not only of the existence of the trending data but also of the likely timing of failure. This enables operations staff to better plan activities such as calibrating and maintaining equipment or to plan training refreshers for operations staff. The characteristic profiles of equipment can also be used in simulations for facilities planning.

For this level of MES functionality, it is essential to have confidence in the data within the system, a strong understanding of the characteristics of proper operational performance of production equipment, and a well-established production support program based on MES reporting.

The only perceived issue with a phased implementation plan is that it takes time to "complete." The assumption is that when a project turns on the production system, the implementation is completed and the project is finished. The activity does moves to more of a support or an "ongoing improvement" model, and there is continued growth in using MES functionality and understanding new ways to utilize an MES. However, the implementation of an MES should never be deemed "completed." Unless there is active planning of MES functionality as part of the continuous improvement management, there is a high likelihood that the utilization of the MES will stagnate.

Implementing Customizations

Another aspect of implementation that must be understood is using and managing customization to add specific capabilities to MES functional solutions. The first part of this "decision-making" must be the recognition that there will be a need for *some* customization. With each plant having unique products being produced on production lines with unique resourcing and sometimes even with unique systems, the line of thought I have used is that it is impossible to design an MES that can support everything without some aspect of customization. Trying to implement an MES without customization will only impact the ability to support the production floor.

However, as discussed in a previous "Real-Life Experience," implementations with a significant amount of customization can be detrimental to the life of the system and to the growth of the company.

A tactic that SI firms have applied in some implementations is using customization as a tool to cover up a lack of understanding of the internal workings of the MES and as a means of driving up consulting fees as part of the implementation. Most MES applications have been designed to provide the capability to link a customization to the system at common locations within the system. Vendors of MES software, while working with organizations like MESA and ISA, recognize the need to handle customizations as a result of variances of business logic (or additional spot analysis requirements) and have provided a means to link customizations without interfering with the logic of the code. However, properly linking customizations of this nature requires an in-depth knowledge of the particular MES. Some SI firms that do not have that in-depth knowledge will implement customizations that completely bypass these link points and override specific business logic within the system. This can lead to the loss of key capabilities in the main system and significantly increase the consulting cost from the SI firm.

There are a few general rules that CoEs should use as guidelines for customization:

- Customizations should never be used to solve a problem that is rooted in poor process management.

- Customizations should be to augment MES capability, not replace it.

- Any customization should provide a means to control the use of the customization via configuration and not via changes in coding.

- Any exception handling needed in the customization must be tied to the MES application's normal error-handling routines.

General Rules of System Interfacing

In the earlier days of the MES industry *(OMG, that makes me sound old)*, there was a considerable amount of activity regarding integration—integration of enterprise systems and integration to equipment on the floor. The popular saying at the time, "Visibility from the shop floor to the top floor," promised to provide the same information to the executives on the top floor that the system provided to the shop floor operators. In each case, the project would design a customized interface to enable data from one system to get into another. Having all this integration was great for getting information from one system to the next. However, this also created a problem with data integrity across the fully integrated system. After the data was transferred from one system to another, because of poor process management at the planning and administration level and a lack of defined ownership of the data, many companies allowed the new data in the receiving system to be edited by local operations staff at the plant level (or vice versa) without ensuring the changes got back to the originating system. This discrepancy in data led to considerable confusion as to which data was correct and finger-pointing as to who messed things up.

When interfaces are set up, the CoE must also ensure there is an understanding of how the data in the receiving system is going to be used and, if it can be changed, how it may change and who owns the changes. This requires establishing *data vaults* to ensure which system is the author (and holder) of that specific data and which systems may need access to that data (which systems must be interfaced). If the company is using an SOA for interfacing, normalizing the data at the SOA level makes ownership and updates easier. However, if not using normalized data, the company must rely on defined procedures and intelligent interfacing to ensure that only one system (and likely department) becomes the "source of data" and that any receiving systems (or departments) maintain the integrity of that data.

The other part of maintaining data integrity is recognizing that there will be exceptions. There will be times when changes must be made outside of the normal procedures, and there must be a means to reconcile these exceptions. Without the reconciliation, the integrity of data and the reporting from that data will degrade over time. For example, there is an ISO-9000 requirement to regularly "audit" processes and the management of continuous improvement. As part of the audit process, the CoE should include auditing the data across systems as well.

Maintaining the System

In an effort to reduce variation in production, a few activities are performed. To maintain inventory accuracy, companies engage in "cycle counting" to check the accuracy of physical inventory against the reported inventory count in the system. To ensure product quality, production equipment undergoes regular maintenance and calibrations. In the same light, MES must also be maintained. The databases must be checked for lost records or other anomalies. Servers must be evaluated to ensure that they are operating correctly, and, of course, there are regular maintenance installs to fix programming bugs. Any IT person can tell you about these things. However, is the system still providing the best support to production? MES vendors regularly issue upgrades to their applications to provide major fixes to operational functionality and to introduce new functionality. It is important for the CoE to also review the current usage of an MES (including any customizations that were implemented) with the new functionality that is being introduced. In some cases, the company may have introduced a mitigation tactic to control problems that an MES application currently cannot correct. Or an upgrade may include new functionality that is meant to replace an older part of the system. As a normal part of IT maintenance, the servers, PCs, and networks are also maintained and upgraded. Any changes in these factors should also be reviewed by the CoE.

Summary

Whether a company is implementing an MES or a CoE for manufacturing operations, having a full understanding of the model of the activities presented in the ISA-95 standards and the skills that are required by the manufacturing floor to support these activities can go a long way to support operations management. The ISA-95 Series of Standards (Parts 1 through 5) was developed specifically for this purpose. In this chapter, I took a deeper dive into the ISA-95 standards and covered a broad range of issues that will be of concern for anyone considering implementing an MES. I also covered criteria that should be examined to understand how an MES relates to a company's manufacturing model, other criteria related to technology used by an MES, and concerns related to maintaining the MES environment.

6

MOM and the Functionality of an MES

Chapter 4 touched on some of the important functionality of an MES. In this chapter, I take a deep dive into MES operational functionality. I also discuss the differences between MES designed for supporting process manufacturing and MES designed for supporting discrete manufacturing. Later in the chapter, I offer some details regarding integrating an MES to the enterprise level and to shop floor equipment.

MES for Process and Discrete Manufacturing

There are two primary reasons for the differences in MES for process manufacturing versus discrete manufacturing. The first is the relationship between the material to make the product and the conversion process used to manufacture the product. The second is the concept of cost savings in manufacturing in large quantities and how those quantities are reflected in the production process.

These differences drive differences in the type of equipment used and in managing production execution, and, therefore, the design of the MES capabilities that support these manufacturing models.

Some industries (e.g., meat production and semiconductor) have manufacturing models that may create variations in the examples provided in this chapter, but those industries will not be covered in this book.

Process Manufacturing

The two concepts outlined in the previous section, for the most part, play out in process manufacturing as taking a large quantity of a material (or in pharmaceuticals, a large quantity of a mix of materials) and processing that large quantity to create a bulk quantity of a product, then dividing the bulk quantity into smaller segments of the same material (or mix) that is sold. For example, if a pharmaceuticals manufacturer attempted to mix the required materials in a quantity small enough to create a single bottle of vitamins, the cost of material handling in such a small quantity would make production costs far too expensive. (Imagine trying to mix a bottle of materials in quantities measuring in milligrams.) Mixing in larger quantities and then separating the quantity needed for each vitamin pill is far less expensive and easier to control. This concept holds true for all products made via process manufacturing.

Figure 6-1 presents an overview of a food processing line that is similar in structure to many process manufacturing models. Like the pharmaceuticals example, the raw materials are processed and mixed in large volumes by equipment designed for that purpose (tomatoes are processed and mixed with spices to create a slurry of tomato sauce). As it would be highly impractical to process a single tomato at a time, all equipment in the line is designed to process material in large batches (in some cases, in a continuous flow of product).

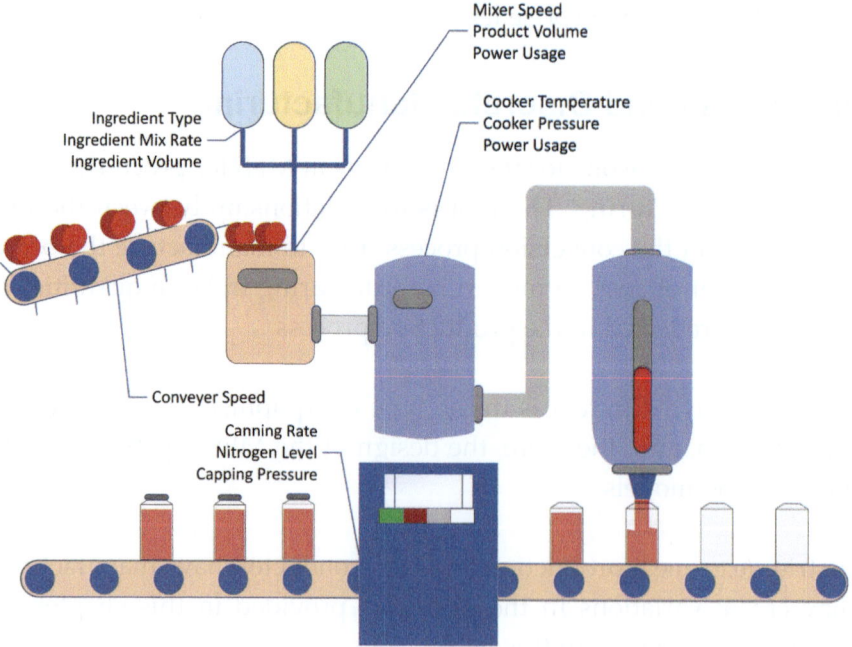

Figure 6-1. Overview of manufacturing in a food production process model.

With this process, it becomes impossible to determine which process characteristics that were affected by a single tomato or to identify a specific activity for a single bottle of tomato sauce. After processing, the large batches are dispensed into smaller units (bottles of tomato sauce) depending on the production order requirements, with each of the smaller units having pretty much the same characteristics. In this kind of manufacturing environment, the same batch of bulk material (the vat of tomato slurry) may be used to create end products in various sizes (e.g., 32-oz bottles or 120-oz bottles), and the quantity of each production unit size is controlled by its own production order. In this production model, the batch of bulk material is processed by a production line that has been set and dedicated to the batch. As a result, the batch ID of the bulk material is an important data element, as is the line that produced the bulk material. Because this information is needed for traceability and is recorded at the bulk material level, the smaller containers of the end product would be labeled with the batch ID or lot number of the original bulk material (the tomato slurry). As all the bulk material was processed by the same equipment and followed (more or less) the same process, the parameters for equipment operations are important over the processing of the entire batch. This context of equipment operations may be the same for multiple production orders, or any single order may have end-product containers (jars of tomato sauce) from different bulk batches. If some parameters of the process are out of specification during the bulk material processing, that is frequently a reason to scrap the entire batch of bulk material because separation of "good bulk material" from "bad bulk material" may be impossible.

So, how does this model of manufacturing affect the design of an MES application?

MES in Process Manufacturing

In process manufacturing, an important factor for consistent product is to ensure that the equipment that processes the large volumes of material operates repeatably. Whether the line is producing 32-oz bottles or 120-oz bottles, it is the relationship of production parameters to the bulk material that is important. As these pieces of equipment will sometimes perform a single operation for a relatively long period (possibly hours or days), the critical relationships are the operating parameters of equipment to the expected recipe parameters with little regard for the particular production order or the format of the end product to be produced (which size of bottles of tomato sauce). The recipe will define the specific equipment to be used, the parameters of the equipment to be set, and the volume of material (ingredients) to make a specific volume of the bulk product. In some cases, equipment may be designed to perform multiple operations in the recipe to reduce handling and moving bulk product from one piece of equipment to the next. Conveyor systems (or product piping) are commonly used to move large volumes of bulk product.

As a result of referencing back to the bulk product, the data structures for MES that support process manufacturing tend to focus on the recipe for the bulk product to be made and the relationship of the specific equipment parameters during the time of processing to the batch identifier of the bulk product. Identifying the production orders that were produced is frequently a secondary data tracking element to the recipe and equipment operating parameters. In addition, identifying individual bottles of the packaged product is limited to the bulk product batch identifier, if the bottles are identified at all. This, of course, is reflected in the database structure for an MES and in the functionality the MES provides. During a production run, Quality will perform several inspections of product samples from the bulk product (sample testing where the results of the sample test represent the results of the entire batch of product), as well as testing equipment and process parameters, all to ensure that the product is acceptable.

Without an MES, all the data would be tested and collected manually and then frequently stored on hard copy or in a customized distributed control system (DCS) at a significant cost for the specialized function. When there is a problem, it is noted and documented, and the batch of material affected would likely be scrapped. Although manufacturing engineers sometimes look for anomalies in the recorded parameters within the historians, without a link between when the inspection sample was taken, when the bulk product was processed through the equipment, and which specific batches of ingredients were used, it is usually difficult to draw a conclusion as to what caused the problem. When the operating parameters are recorded by an MES in addition to all the processing activity timelines for the bulk product, it becomes easier to identify possible causes and measure the correlation between activities, operating parameters, and sampling concerns. Using an MES also helps enable automated sampling of bulk products and directly linking the sampling results to the complete history of bulk material processing, which in turn provides greater capability to analyze the cause of processing anomalies.

Discrete Manufacturing

In discrete manufacturing, the conversion goes in the opposite direction from process manufacturing. Many smaller independent components (materials) are processed and *transformed* into a single, larger end product. As a result, discrete manufacturing processes are created to process the individual materials needed for a single production unit and to manufacture individual units repeatedly in large quantities that (theoretically) are exactly the same or at least very similar.

Figure 6-2 provides an example of a generic discrete manufacturing model. In this context, manufacturing is to create a single product by adding materials together in some

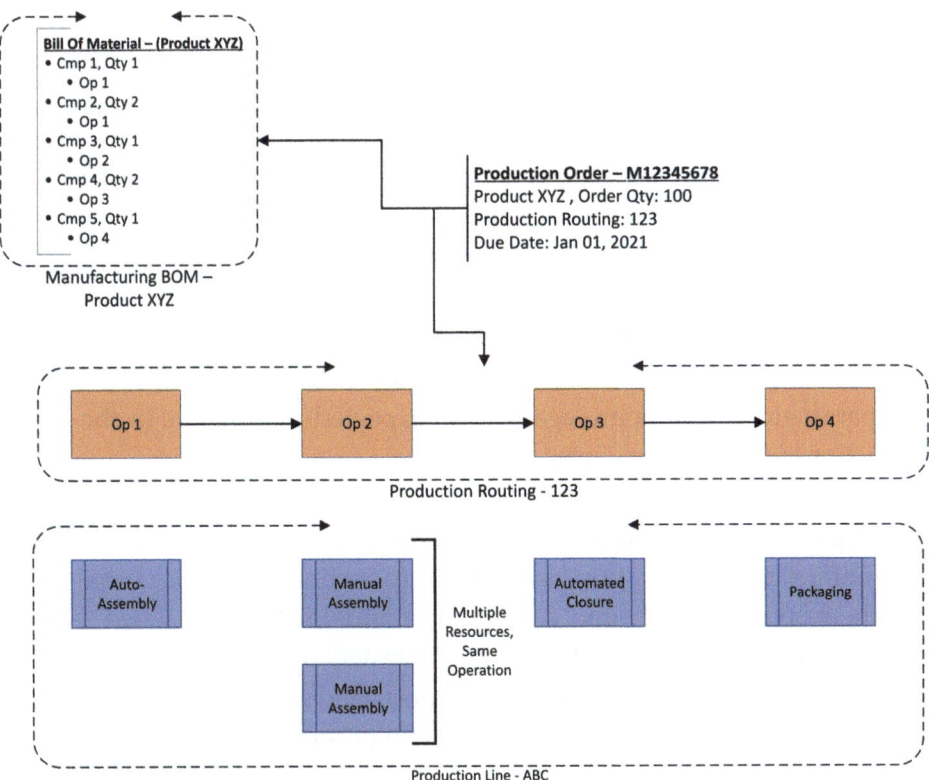

Figure 6-2. Overview of manufacturing in the discrete model.

form of assembly. (Multiple component materials are "assembled" together to produce a single production unit.) A production line would then repeat the assembly process for each production unit as many times as required to manufacture the number of production units that are required by a production order. In some cases of hybrid (process and discrete) manufacturing, segments of the full process may fabricate many units of a subassembly in a process manufacturing model that are then used later in a discrete manufacturing process to build units of the "end product," which are then sold as top-level assemblies. In an actual discrete manufacturing environment, each operation (*process segment* in ISA-95 terms) is defined to process one production unit many times. As a result, the full process will create various subassemblies (or *build levels*) as the "tracked production unit" is moved through the full process.

In discrete manufacturing, there are frequently multiple resources that can perform each operation (or process segment) and, as a result, operations are defined to use a "class of resource" (a resource type), with the actual resource used to be determined at the point of use. In discrete manufacturing models, the process should be defined separately from the production line (or cell). This provides flexibility in the planning for resource utilization.

The production order in discrete manufacturing is to build a specific quantity, of a specific product ID, using a specific bill of material. (In some manufacturing environments, the use of a set time-period, typically a month, is used instead of a set quantity.) The manufacturing bill of material (mBOM) will define which material (components) is to be consumed (assembled) at specific operations. The details of what constitutes a "defined material" may vary, with many definitions being as detailed as including the batch ID or the serial number of the component and the serial number of the product it is to be assembled in. In recent years (particularly in Industry 4.0), the acceptable parameters of each operation for each class of resource are also provided in the mBOM, and the systems that monitor production (MES) collect the actual runtime operating parameters to compare against the expected parameters in the mBOM for each production unit being produced.

If operating parameters at the resource level were to drift out of specification, this might only affect the limited number of the production units that were being processed by the resource at that time, causing individual units to be listed as *nonconformances*. As a result, production units are usually tested and inspected at the individual unit level. Each individual unit may (or may not) need to be processed in a repair or rework operation to bring it back into specification or scrapped without affecting other production units in an order.

MES in Discrete Manufacturing

Depending on the product being made and the processes that have been defined, each production unit may have a unique path through the production process (including repair/rework steps) and require unique identification to track its movement. As a result, MES implementations for discrete manufacturing also have serialization capabilities (serial numbers) to identify and track each production unit and to possibly alter the serial number as required by the process without losing the history of that specific production unit. An MES designed for discrete manufacturing will have the production order number and/or the unit serial number as the primary data object, with all other data linked to one of these two data objects during runtime operations. This will provide visibility to all activity that has been applied to each individual production unit (serial or batch).

Choosing an MES for Process or Discrete Manufacturing

Because there are significant differences in the data and the functionality offerings between an MES for process manufacturing and one for discrete manufacturing, it is important to understand the manufacturing models that are prominent within each company. In companies that are vertically aligned (multiple plants produce subassemblies that are assembled into the end product, as covered in Chapter 5), it is sometimes difficult to isolate a specific manufacturing model. To further complicate this determination, because of the production volume for a single product, some plants manufacture discrete

> It is important in this case to remember that the company is a *manufacturing* company and not an IT company, and the priority should be to optimize the manufacturing.

products but track and manage production in a process manufacturing manner by processing a larger volume of simple products (one or two-step processes) and managing quality and nonconformance dispositioning at the batch level. When selecting an MES, it is important to choose a system that supports the broadest number of manufacturing models within a company. Some companies make the mistake of assuming that it is best to have only one MES application for the entire company. (Projects managed by the IT department usually makes this type of decision.) If the system integration is designed properly (see the later sections on integration), it may be more effective for company management (and effective manufacturing management) to support multiple MES. It is important in this case to remember that the company is a *manufacturing* company and not an IT company, and the priority should be to optimize the manufacturing; the effective manufacturing of products must always be the number one goal. If only one MES application is needed (a management decision), it is usually easier to expand the use of a discrete MES into a process manufacturing environment than to expand a process MES into a discrete manufacturing environment.

The Functionality of an MES

With an understanding of the MOM model that an MES supports, it is now possible to look deeper into MES functionality and understand how these functionalities *should* be used to support MOM.

In this section, I examine the different functionalities provided by an MES and provide a deeper explanation of how these functionalities interact in supporting MOM. The first aspect of MES functionality that must be understood is real-time transaction processing. Figure 6-3 shows the MES background activity for every transaction performed by a production operator.

Whether a production operator starts a transaction via a computer screen or a machine interface, the background activity is similar. The system will accept the transaction request from the operator, validate the transaction against several criteria (explained later in this chapter), and, only after all validations are complete, perform the transaction and send a reply to the operator (or equipment) acknowledging a *completed transaction*. In MES transaction processing, the operator (or resource) requesting the

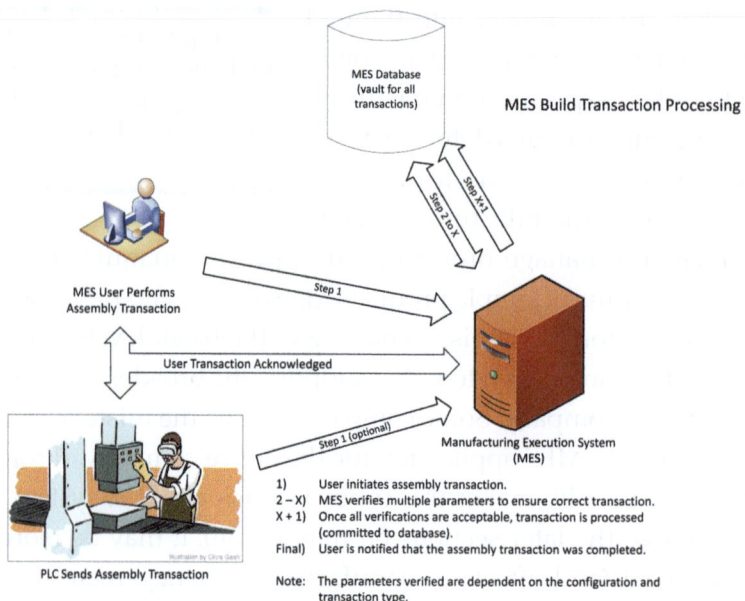

Figure 6-3. MES transaction processing.

transaction to be performed will be required to wait for a response before continuing to the next transaction. With most enterprise applications (systems at ISA-95 Level 4), the transaction receipt will be acknowledged, and the transaction will be placed in a queue to be processed sometime later. The requesting operator is not expected to wait for a response from the system before continuing with the next transaction.

In the following sections, I provide a more detailed explanation of the functionality of the MES. These are the MES functions that will be covered:

- Track and trace

- Production order execution management

- Process validation management

- Resource validation management

- Quality execution management

- Labor utilization management

- Line-side stocking inventory management

- Data collection

- Data validation

- Performance analytics

- Work instruction presentation management

Track and Trace

Track and trace, probably one of the most fundamental functions of an MES, is also one of the most widely used. The premise of all track-and-trace functionality is that each production unit is identified to the system, each unit has a record of what actions should be taken on it, and any action taken on any production unit is recorded against it. Having the full record of all activity expected by a process, verifying all transactions requested on the production unit, and recording all action taken against a production unit in real time enables the system to guide production operators in their work and immediately identify missteps back to the operator. It also enables production and quality operators to trace process anomalies back to their origins, which helps determine the root causes of those anomalies. In addition, identifying production units can help resolve greater quality issues that result in warranty claims. There are a few issues with production unit identification, the first of which is answering the question, what is a production unit? The answer depends on the industry and a company's product structure. Also note that the "real-time" aspect of an MES (addressed later in this chapter) provides the ability to record, verify, and update the current status of a production unit for every transaction performed on the unit as the transaction is being performed. After the unit identifier has been set, this data can be tracked during the other assembly operations, such as when the unit is assembled into a higher-level product. This is referred to as *tracking the genealogy of a product* (tracking all the material and components used during manufacturing by the unit identifier). The end result is a record of all activity that was applied to any specific production unit and the complete makeup of all material that was used to make that production unit. In discrete manufacturing, this entire record is sometimes called a *device history report* (DHR).

More information on production unit identification for process and discrete manufacturing is provided in Chapter 8.

Production Order Execution Management

In a manufacturing plant, a planner works out a schedule for production orders based on customer needs, available material, and available resources. The planner keeps track of when materials and resources are due to be available and releases the order to the production floor just in time for the order to be completed and shipped to the customer, and everyone is happy. At least, this is the way the process *should* work.

The issue is that production execution management is easy only in a perfect (or at least near perfect) world. When a production order is released to the production floor, there are still a few things that can go awry, and being aware of what is happening on the production floor in real time can make all the difference in the world (*or at least in the plant*). There might be equipment failures, resource availability changes (e.g., equipment calibration took longer than expected or a production operator got sick or was injured), or a quality issue with a batch of material that was discovered during production, making it unavailable when needed. There are things that can go wrong that must be monitored in real time. Even something as small as a production line running a few units per hour slower (e.g., due to someone being trained on the job during production) can make a big difference to a production delivery schedule. If we are talking about monitoring a piece of equipment, the programmable logic controller (PLC) system will likely be able to provide that warning as well. However, determining how the slowdown of one piece of equipment is affecting the remainder of the production line is out of the scope of most PLC systems. Not only is this *in scope* for an MES, but the MES is also capable of identifying particular orders and products that may be affected by these events, enabling the supervisory staff to react to them and make adjustments to the line (or production order schedule) to compensate. The sooner these notifications become available, the easier it is to manage such events.

It is the planner's function to schedule production orders to the floor within a specific time frame (usually down to the shift level), but with the events that are happening on the floor, it is difficult (some would say impossible) for the planner to keep on top of all the schedules. Therefore, many MES also provide the ability to *dispatch* (assign a production order to a specific production line or operator) a production order at the floor level. (This is usually done by the line's supervisor.) This is an important function, not only because of the ad hoc events that are happening but also to manage the efficiency of line changeovers and material restocking.

Process Validation Management

In Chapter 3, I addressed the importance of repeatability for manufacturing. When the process is defined, the system can lead the production operator (and, therefore, the movement of work in process) to the next step in the process. But how does an MES *validate* a process? This function uses a *detailed master data record* of the process *to be* followed that is configured in the MES before production starts (either as part of implementing the MES or as an update prior to a production run). This master data record may have been provided by a product lifecycle management (PLM) system and imported into the MES, or it may have been created in the MES. With the process master data, the MES knows what the process should be, and uses the real-time track and trace function to ensure that each step (operation or task within an operation) started by a production operator (or automated equipment) is correct. So, whether it is a production operator or an automated manufacturing system, the first step for the operator (or equipment) is to signal the MES that they are about to start a particular operation with a specific production unit (batch ID or serial number). The MES will then evaluate where the production unit is in the process (tracked by the history of what has been completed) and the next step in the process according to the master data and then compare the signal from the operator to the master data. If the operation/step in the signal data matches the master data for that specific production unit, the MES will *start* the operation, reply to the production operator that the *start* was successful, and then change the screen for the production operator to perform the production operation required. If, however, the operation defined in the original operator's signal does not match the correct next operation for that production unit, the MES will not initiate the operation and will reply to the production operator regarding the discrepancy (provide an error message). The production operator *will not* be able to perform the work unless the discrepancy is fixed. The production unit identifier must be changed to a unit that is actually at that operation, or the production unit must be moved to the correct operation to match the correct operation in the master data. As part of the process definition, the details of the sequence of steps within each operation are also defined, monitored, and validated. (Think of validating the assembly steps of each of several components being assembled in a single operation.)

This type of *validation* is performed by the MES for every transaction that is initiated. As a result, the more detailed the master data is within an MES, the more transactions that can/will be validated. This also indicates the need for the level of real time processing that is handled by MES.

Resource Validation Management

Similar to process validation, the MES master data is configured to include all equipment and/or tools used during production and a description of what type of resource each piece of equipment or tool is. In most MES, a resource can be defined as multiple resource types, depending on the level of control needed by the production floor. At the same time, the master data record for the operation (explained earlier in process validation management) is defined with data that describes the *type* of resource that can perform the operation. In this case, there should only be one type of resource allowed. While the MES is validating the process, it also validates the resource identifier in the signal from the operator. In addition, the system will validate the resource type the operator is using to ensure that it matches the master data configuration, with similar results as the process validation.

The level of detail that is validated before a transaction is performed depends on how the master data configuration has been set up in the system and if the MES has access to the equipment data that must be validated against the master data. Included in the potential configuration of an MES are the equipment settings, the usage history, and the runtime operating parameters of the equipment or tools, any of which can be used to identify the *state* of the resource. If the MES configuration defines what the operating conditions of any resource or tools on the production floor are supposed to be, the system can be used to validate those conditions before any transaction in the system is performed.

An additional validation that can be included in the MES is to ensure that any particular resource is available for use (and has been allocated) for an operation. The MES can check for criteria such as whether the resource has been scheduled and whether another operator is currently using that resource, or perform other checks regarding the resource's current state or fit-for-use.

Quality Execution Management

Quality execution management is primarily about ensuring that the processes for each production unit are maintained as being *in control* or each production unit is tested or inspected as required, and the results of these tests and inspections are properly handled (or dispositioned) for each production unit. Depending on the industry, the aspects of quality management can be handled very differently, and the details must be carefully considered.

This MES function must be able to define the test or inspection criteria to be evaluated at the operations level or to define when and how a process is to be monitored (e.g., statistical process control criteria of variables to be tested, access to the data in real time, sample rates and type of sampling). It must also give the production operator the ability to interact with the quality execution process. In addition, quality execution management includes reacting to nonconformance issues: identifying nonconforming units, quarantining these units, and ensuring full disposition to a final state of return to production or scrapping, depending on the results of the disposition process. This is an area where there can be considerable differences in the available functionality.

In addition to monitoring the test or inspection operations, the MES can be configured to monitor the available training and/or certification of production operators. This function ensures that operators are trained to perform the defined work and also tracks when these operators are due for retraining. If there is a mismatch in the required certification to perform work or if the certification has expired, the MES will prevent the operator from performing that function.

Labor Utilization Management

In many manufacturing industries, labor tracking is quite simple, and frequently a labor allocation is simply assigned to a production segment at the enterprise resource planning (ERP) level. In this situation, a proportion of labor cost is automatically assigned any time a unit of work is reported as being complete. In this environment, there is little advantage to using an MES to perform labor tracking.

In some industries, however (e.g., aerospace), a highly skilled manufacturing operator may be required to work on many production orders during the day and may have specific labor charge codes to use for different work within an order. Because the MES is *aware* of when that manufacturing operator is working on something and *what* the operator is working on, it can be configured to automatically assign labor charges to correct labor codes specifically for that operator. Provided the MES has been configured properly (including all user interfaces), the system can identify all touch times for all operators at any time.

Line-Side Stocking Inventory Management

In most companies, the primary application for managing inventory is the ERP system, although supply chain management systems are sometimes used. The ERP system is used to manage and control the movement of inventory for a large proportion of operations activity, including, in most cases, the consumption of inventory into manufactured products. When a quantity of production units passes an ERP reporting point on its router, the material for the segment of the manufacturing line is consumed and costed. However, in many situations, the ERP reporting point covers the consumption of multiple production line operations. In these situations, that material will have been physically consumed but will not be reported until a few operations down the line. Depending on the line rates and nonconformance handling, this difference can be anywhere from minutes to days. The MES solution to this problem is the line-side stocking (LSS) functionality for inventory at workstation locations. In this scenario, the ERP system transfers material not just to the floor but to a specific location on the floor (and transfers the record to the MES, which manages the actual consumption). Material is marked as "allocated" in ERP when it is transferred to the MES. Any movement of material from one LSS location to another is managed at the MES level, and it is the MES that defines when the material is consumed and signals ERP. This type of material management in an MES is important for high-value material and consignment material management.

LSS material management is important to the function of material supermarkets that feed Kanban or kitting functions. It is also required for the effective use of automated inventory movement using automated guided vehicles.

Data Collection, Validation, and Performance Analytics

A significant part of any MES is the function of real-time data collection. This, however, is not a singular function. Data collection spans almost the entire breadth of MES functionality and is an important part of all the MES functions that have been discussed in this chapter. Each of these functions has the means to define where, when, and how (manual entry, scanned, or automated interface) data is to be collected within a process. After that data is collected, it becomes immediately available to any other function in the MES. An additional part of data collection is the validation of "correct data." Depending on the configuration of the data collection function, most collection parameters provide the ability to define the limits and/or type of data that is to be collected. If a data collection function is defined with these limits, the system can also define how to respond to limit violations (rejection, process stoppage, warning, etc.). Because data that has been collected is available immediately, the MES is the ideal system for processing performance analytics to be used for event monitoring, trend analysis, and dashboard reporting for all areas of the previously discussed MES functions for production management.

Work Instruction Presentation Management

Another scope of the MES function that comes from the real-time capability is identifying which specific work instruction is needed for a specific step in the process. Because each step in the process is identified and verified before the step is performed, the MES can also fetch and present the correct work instruction that the operator needs at that particular moment. Sometimes these work instructions may be static (the same instruction each time), but when this function is combined with the data that is available via real-time data collection and analysis, a dynamic instruction related to a specific range of data that was collected moments before can be provided to an operator.

Integrating an MES Using ISA-95

The first important aspect of the ISA-95 standards to remember when designing the MES integration is that the standard *does not define which IT system is used to support any particular section or level of the standard.* In this context, it is up to the company (or at least the system integrator) to define which IT system is to be used to support a specific MOM function. That being said, the ISA-95 standards were developed in conjunction with MES vendors, and, as a result, the vast majority of specific MOM activity is supported by MES applications—so much so that Level 3 in ISA-95 (where most of the MOM activity resides) is frequently (although officially incorrectly) referred to as the *MES layer*.

The activity models for each of the management sections of the standard (remember the four pillars) and the data models used to support these activities also define activities (and data models) that are external to MOM. The activities that are external to MOM, which reside at what is defined as Level 4 of the activity hierarchy, are viewed as coming from the "enterprise level," and therefore, when viewed from an IT perspective, will come from either ERP or PLM depending on the enterprise IT stack (the list of applications available in an enterprise). Figure 6-4 shows the ISA-95 activities according to the levels of hierarchy.

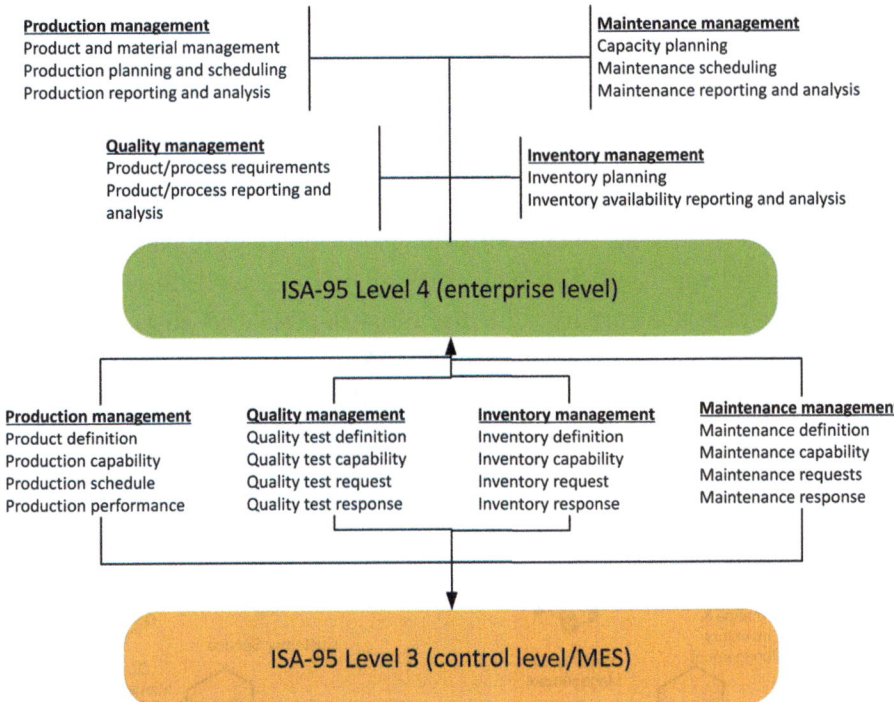

Figure 6-4. ISA-95 activity hierarchy Levels 3 and 4.

In addition, the standard specifies activities and data models external to MOM that reside at levels lower than Level 3. These activities (and data) are viewed as coming from equipment or people on the manufacturing floor and include controls systems such as PLCs and test equipment. In this perspective, the activity is performed at Level 1 or 2, and the data is then provided to Level 3 to ensure a holistic MOM context.

In the next couple of sections, I explore the levels in ISA-95 and integrating the systems to support the standard.

Integration with the Enterprise

From the perspective of IT systems, there are some best practices integrators should be aware of. When designing the interfaces between the enterprise systems and the MES

Depending on the type of systems used, many *knowledgeable* integration specialists recommend the use of a service-oriented application (SOA) to build the interface architecture.

to support MOM, it is important to recognize that the company's processes (and maybe even the primary business model itself) are going to change over time. As a result, the application stack (the set of applications that support the business) is likely going to change over time as well. Some of the applications in the stack will be upgraded or may be replaced. As changes in the application stack are to be expected, it is important to create an interface structure that has the capability to change (relatively) easily as well.

Enterprise Integration Using an SOA

Depending on the type of systems used (more on this later), many *knowledgeable* integration specialists recommend the use of a service-oriented application (SOA) to build the interface architecture. As shown in Figure 6-5, using a well-designed SOA implementation will provide a company with the most flexible and cost-effective application stack possible. By using a well-designed SOA infrastructure, the company could upgrade or replace any of the applications within the stack with the least cost over the life of the enterprise. A full description of an SOA is not within the scope of this book;

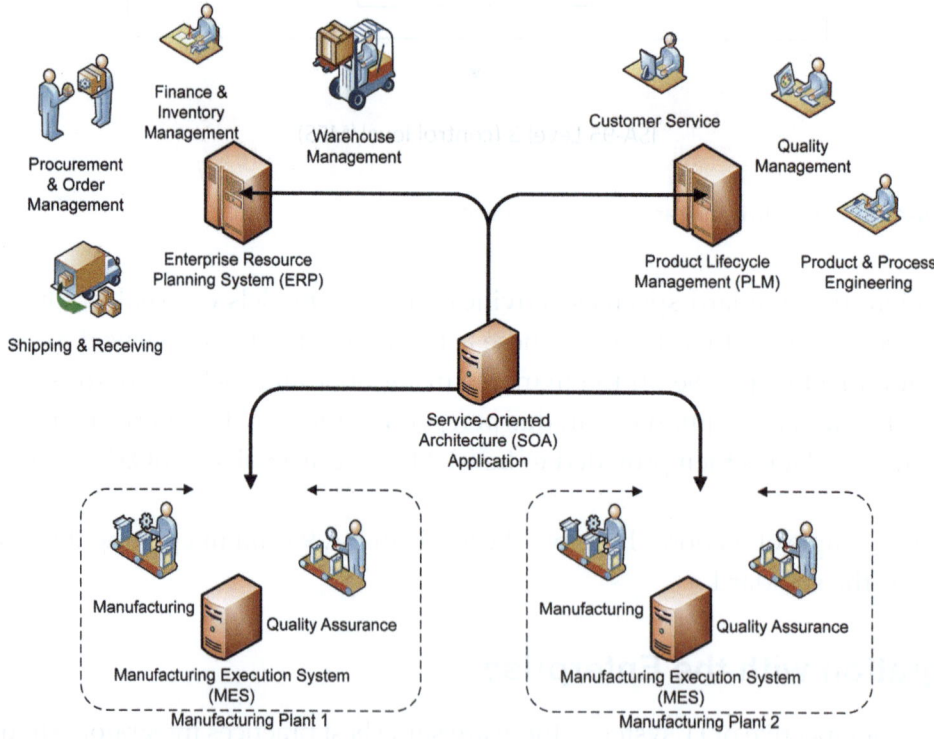

Figure 6-5. Enterprise architecture using an SOA.

this discussion has been technically simplified to relate some of the more important points for MOM support.

After deciding to use an SOA infrastructure, the next decision is whether to use what I refer to in this book (to simplify discussion) as a *thin-layer* SOA or a *fat-layer* SOA. A *thin-layer* SOA translates a data block in one application's data structure directly to the structure of the next application before transferring it. A *fat-layer* SOA translates a data block to a *normalized* data structure before transferring it to the next application.

Figure 6-6 represents the thin-layer SOA implementation that is connecting PLM, ERP, and document management (DM). In this implementation, a data block from any application is *picked up* by the SOA application. The system is designed to recognize the structure of the data and parse the data to prepare for sending the data block to the next system. If there are multiple systems to receive the data, the SOA will structure the data for a particular system and perform the transfer to each receiving system independently. So, a *bill of material* (BOM) data block coming from PLM would be parsed to send a planning BOM to ERP (e.g., not including equipment setup data) and then parse the same data from PLM to send the mBOM to the MES (including equipment setup data). After the data transmission is completed, all data in the SOA layer for that particular transfer is deleted. The SOA application is not expected to retain that data for a long period. However, there should always be a log of which actions were performed for communication analysis.

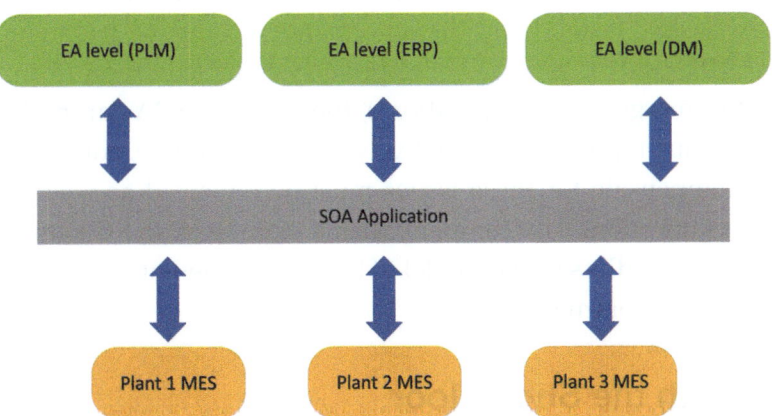

Figure 6-6. Thin-layer SOA.

Figure 6-7 represents a fat-layer SOA used by companies that typically have a few regional offices or retail centers. In this SOA implementation, any data received in the SOA layer must first be translated to a normalized structure before it is translated and sent to the next system. The data that is "normalized" is maintained long-term as a centralized record of all data that is interfaced with other systems, and data dictionaries are used to ensure that the context of the normalized data is maintained throughout the company's IT systems and with any upgrades in any of the IT systems.

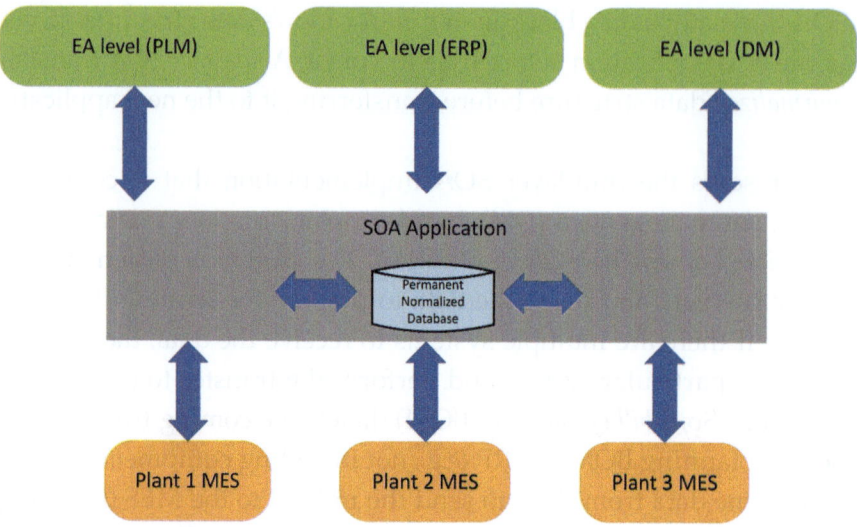

Figure 6-7. Fat-layer SOA.

Each of these implementation structures has advantages and disadvantages. For example, the thin-layer SOA is easier to implement and maintain. It does not require a large database at the SOA level, and, given equal technology, the thin-layer SOA transfers data to the next system faster. If, on the other hand, a company has many plants that have different MES and/or multiple enterprise systems (common in industries that engage in a lot of mergers and acquisitions), the fat-layer SOA can make integrating these companies much quicker and relatively easier. Of course, communication strategies that can compensate for these issues must be reviewed as well for the overall strategy of communication. Any company that is looking at an SOA must evaluate its particular needs. Regardless of the model of SOA implementation, the use of an SOA, in general, is highly recommended.

Integration with the Shop Floor

When a manufacturing company has gained experience with an MES, it will also develop a significant requirement to automate the data collection and improve the reliability of shop floor data in general. In designing the interfaces to the shop floor, it is important to ensure proper demarcation of control and define which system will be the primary system of control for processing and storing (or vault) the data. Figure 6-8

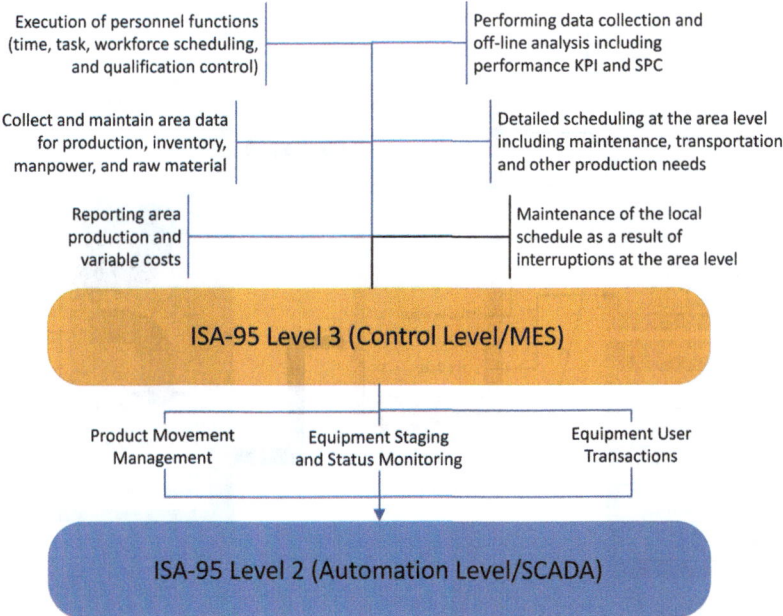

Figure 6-8. Demarcation of activities for ISA-95 between Level 2 and Level 3.

provides a representation of the activities that are performed at Level 3 and the data transactions that would be communicated between Level 2 and Level 3.

Proper interface design, however, requires defining direct interfaces from manufacturing and testing equipment on the floor to ISA-95 Level 3 (assuming the MES). In many cases, MES implementations have created highly customized interfaces to the equipment layer (Levels 1 and 2 of the ISA-95 hierarchy of activities). Each piece of equipment is then independently interfaced with the MES. However, these interfaces are expensive to create and difficult to maintain.

Controls engineering has created some best practices that help minimize the cost and maintenance of these interfaces.

Typical Interfacing between Levels 2 and 3

In setting up an interface between Levels 2 and 3, there are some issues that developers and controls engineers are generally aware of. Figure 6-9 presents a simplified interface for some equipment. In this interface example, the equipment has two operational parameters that are to be communicated to the MES. Many interfaces at this level use an OPC server (Open Platform Communication–Unified Architecture or OLE for Process Control, depending on the age of the standard you reference) to establish a link between the PLC system and any IT system. These interfaces are

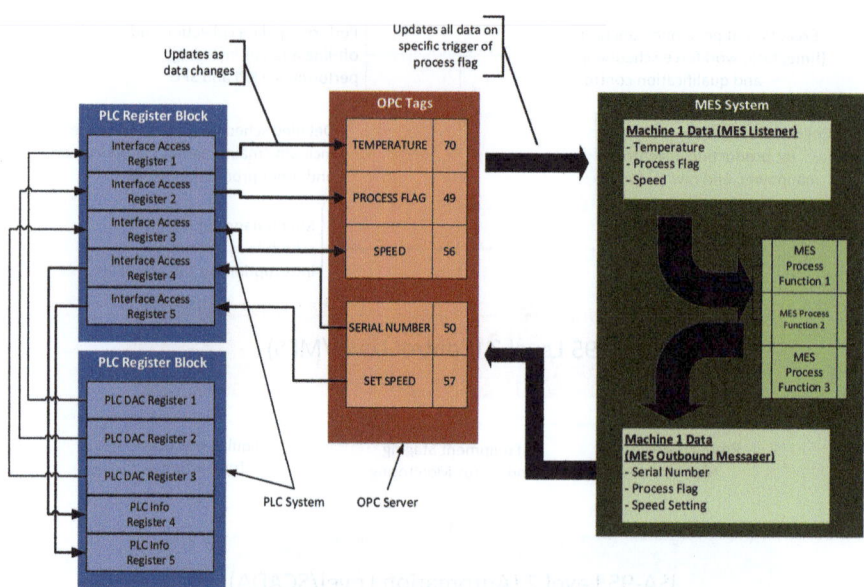

Figure 6-9. Simplified interface for Levels 2 and 3.

designed to obtain data for operating parameters from the equipment or to provide data from the IT system to the PLC system to establish parameter settings and other data used in the equipment's human-machine interface (HMI) functions. These interfaces can be set up to poll for data from the PLCs or can be free running and use a process flag to identify the right timing to trigger a transaction to read the data. If the IT system (in this case, an MES) is receiving data, it will set up what is called a *listener* to wait for data files from the OPC server and then send the data file to other MES functions for processing.

The primary issue with the interfaces, as mentioned earlier, is that they are complicated and highly unique in relation to each piece of equipment. This makes interfacing with equipment expensive and difficult to maintain.

Communication Using Normalization between Levels 2 and 3

To reduce the complexity and cost of interfacing with equipment on the production floor, controls engineers could use a historian to store large blocks of data, as shown in Figure 6-10. Each block would have the same data structure with all the equipment parameters used on the production floor in each structure. If a parameter was not used in a particular equipment data block, the associated parameters with the structure assigned to that equipment would remain as a null entry (not populated). The interface from the historian to the MES would remain the same for all equipment. This makes the interfacing much simpler and much less costly. Data collection

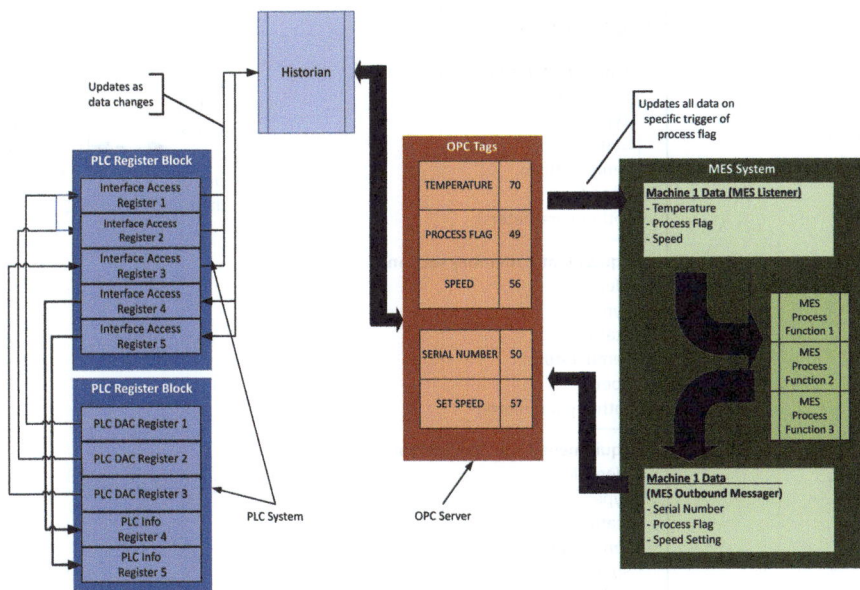

Figure 6-10. Simplified interface using a normalized data structure in a historian.

parameters for specific equipment could be added from the PLC system to the historian as required, provided the parameter to be stored was already available in the normalized structure. If an additional data parameter was needed in the normalized structure, it would be added to all equipment interfaces to keep the communications normalized.

The key function is to normalize the format of data tags before communicating the data to the MES. Figure 6-11 provides a simplified data model of normalized interfacing between Levels 2 and 3. In this interfacing model, a historian is used to hold all the data parameters before communicating to and from the MES. Each piece of equipment is assigned an address sequence in the historian with the same data model used for each piece of equipment (each piece of equipment has a similar data structure and variables assigned to the data model). The PLCs will update the required data variable in the data model from the independent registers in the PLCs. When a variable in the tag is changed (or a set of variables indicating that the historian data is stable), the appropriate flag in the tag is set to initiate a read/write operation with the tag and the IT systems at Level 3. As mentioned earlier, if a particular data variable is not needed for specific equipment, it will always be communicated as a null variable. In this structure, the communication protocol between Levels 2 and 3 will always remain the same for each piece of equipment, significantly simplifying the design and the interface upkeep.

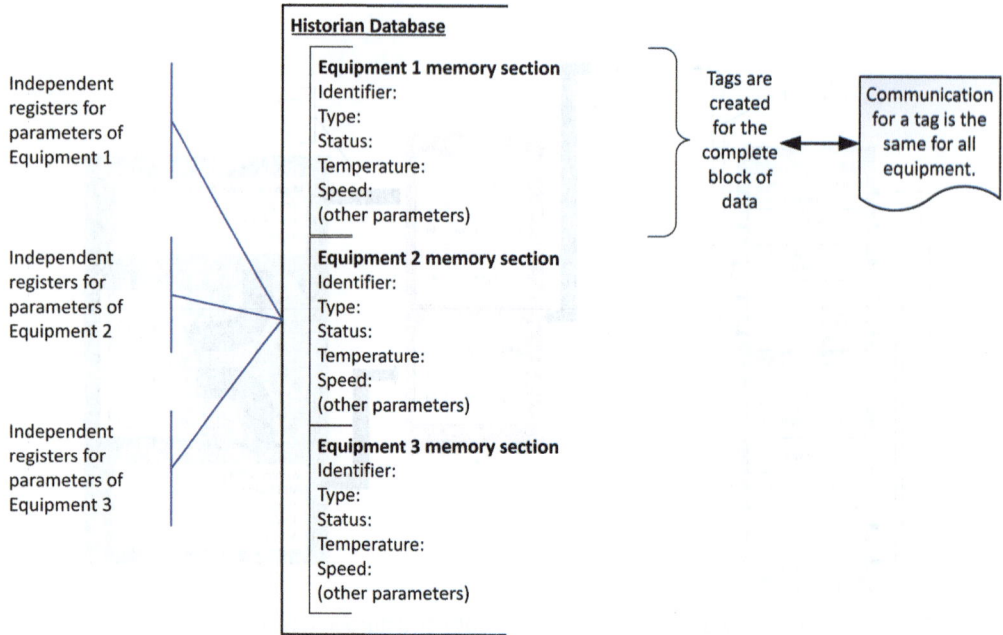

Figure 6-11. Sample historian data for normalizing equipment interfaces for PLC communication.

An additional benefit of this interfacing method is that as long as the tag structure stays the same, data can be made available within the tag as required. If additional variables are needed as part of the tag structure, they can easily be added at the historian level and the data variable to be populated via the PLC is optional.

Communication with the IIoT and MES

Initially mentioned in Chapter 2, recent improvements in communication have made it possible to have PLC devices (sensors, actuators, etc.) connected directly to IT networks using "web services" (web page interfaces to collect or deposit data) or "publish and subscribe" protocols where a device publishes changes in data to a system called a *broker*, the broker notifies all "subscribers" to the change in content, and the subscribing systems (or other devices) collect that data from the broker as required. This is all part of the Industry 4.0 and Smart Manufacturing initiatives. These changes in technology (from a communications point of view) are making it much easier to interface with PLC devices and collect data. Plus, having these devices that are network ready for real-time communication is making much more data available to IT systems like MES. (Identifying which of these technologies is best depends on the specific implementation and is beyond the scope of this book.) The new issue that arises from all this available communication is that a lot of raw data is made available at any time.

There is a need to design and maintain cost-effective interfaces between the equipment and IT systems like MES. If decisions are made in the early stages of equipment design with the scope of interfacing with an MES, establishing communication standards, as presented in this chapter, can help the controls engineering team and the IT team work well together.

Summary

As illustrated in this chapter, manufacturing operations management requires a lot of knowledge to support the manufacturing floor, and the ISA-95 standards were developed to help companies gain an in-depth understanding of their process management and support. The MES provides a great deal of functionality in support of MOM. Some say that an MES is as deep functionally as ERP is broad.

So, what does this have to do with a CoE?

Effective use of an MES within a company depends on a thorough understanding of MOM from the perspective of the ISA-95 standards and a detailed understanding of specific MES applications in support of MOM. The skill set needed for MES/MOM program support is complex and must be managed over time. Developing a sound program of management using the concept of a CoE is key to successfully managing manufacturing as a whole and to the ongoing support of a company's continuous improvement program.

7

The CoE in Strategic Planning and Management

At the beginning of Chapter 1, I provided a brief discussion on sales and operations planning (S&OP), which many manufacturing companies use to align the operations activities with the sales and marketing aspirations. The purpose of S&OP is, of course, to ensure that the sales and marketing activities are being supported by the operations activities (which makes sense, if only more companies did it). In a manufacturing company, however, a wide range of strategic execution activities can be performed in operations, including inventory management and distribution, customer satisfaction improvement, and others. Other than customer satisfaction, I believe that the most significant activities will be related to manufacturing operations.

In this chapter, I review some strategic planning methods from a manufacturing company perspective and present explanations as to why it is important to have a CoE actively engaged in the strategic planning process.

In strategic planning, it is important that the planning be based on the capabilities of day-to-day operations (*business fundamental activities* in the Hoshin Kanri planning methods) and the more significant capability improvement plans (*breakthrough activities* in the Hoshin Kanri planning methods). At the same time, with manufacturing operations, it is necessary to plan day-to-day operations and capability improvement activities for both the manufacturing processes *and* the business processes. Although there is something to be said for initiatives to improve the methods for distributing and managing finished goods inventory, for example, and designing a product with features that are needed by the customer (otherwise, the customer will not be interested in the product), if the planning and improvements in manufacturing operations are not effective, the other initiatives will fail as well. The quality of the product will suffer and/or the cost will be too high, leaving customers with little desire for the product. With that line of thought, some companies have well-designed products but still fail to deliver in terms of well-manufactured products. Quality, availability, and cost of the products are all directly affected by manufacturing.

The issue with the strategic planning process of S&OP is that although it drives reconciling operational capability with Sales and Marketing objectives, it does not define *what to reconcile* very well, and most companies focus on capacity (planned and needed) as the main area for strategic planning. In addition, many companies' strategic plans are frequently stated as financial objectives (this has also been my perspective, albeit based on limited observation), and top management then leaves it to department heads to define how those financial objectives are to be met. Frequently, activities in one departmental plan conflict with those in plans from other departments because each department head tries to drive their own objectives.

A couple of the key concerns about strategic planning and executing these strategic plans are ensuring that the information used is correct and up to date, and that the correct direction has been set in the objectives to be accomplished. Having clear visibility of performance measures "from the shop floor to the top floor" will help ensure that the planning and execution of the right activities have been established from the start.

When strategic planning is taken to the expert level provided by methodologies like Hoshin Kanri and/or balanced scorecard (BSC), the availability of data (and the integrity of that data) becomes even more important. (A further explanation of Hoshin Kanri and BSC is provided later in the chapter.) However, when looking at the strategic planning for the MOM environment, it is important not only to have data but also to have the context of that data understood. This is where using systems like an MES comes into play, and ensuring the context and integrity is the role of the CoE.

Real-Life Experience

Working as a newly appointed manufacturing manager, a colleague of mine (I will call her Jennifer) was participating in her company's planning cycle for the first time. In previous planning cycles for the company, the primary measures used were the cost per unit and total throughput of production. Based on this information, the company had plans to increase capacity via automation (which initially sounded like a good idea). In investigating the situation, Jennifer found that from a purely cycle-time point of view, the production floor should have had no problems meeting planned capacity requirements with its current resources, and she had requested a hold on the decision to automate parts of the production floor. After a little further investigation, Jennifer showed that the production floor currently had a first-pass yield (FPY) of only 60%, and there was no means to determine the typical number of repair cycles needed to get a production unit back into the normal production flow. (The lack of numbers for repair cycles is a common issue for many companies.) Jennifer's presentation to senior managers showed that the issue with the production floor was not capacity but quality. Jennifer then suggested a change in the strategic direction based on stabilizing processes and improving quality based on FPY, a common quality marker for manufacturing. She also suggested running a series of design of experiment (DoE) initiatives on a single line to test this direction.

After taking several months for the initiative to stabilize processes (reducing variation) and several cycles of continuous improvement to bring FPY up to 83%, Jennifer found that the production floor was not only meeting capacity requirements but was also operating at just over 76% utilization. This demonstrated that with current resources, they had some wiggle room to meet current demands. After presenting the results of the DoE initiatives to senior management, the automation plan was put on hold (there may still have been a need in a couple of years), and a new initiative to roll out the process improvement and quality initiatives to the remaining lines was implemented.

The primary point of this real-life example is the importance of visibility and context in data and information for efficiency at the operations level and at the senior management levels for strategic planning.

My experience from years in manufacturing operations management has been that many department leaders and analysts do not understand the significance of ensuring the proper alignment of objectives from the department level to the mandated objectives of the senior manager's strategic planning. I have had colleagues ignore strategic directions from senior managers two to three levels higher and start on initiatives without considering that they were in direct conflict with the company strategy. Their line of thought was that it was (in their opinion) "a better direction to take." The issue with this mismatch is that the senior management may or may not have the right strategic direction, but they are responsible (and accountable) to stockholders for the company's direction. By not following the corporate strategy, my colleagues were causing confusion in the results of any reporting measures for the strategy (like any variance in process, they were adding noise to the reporting results) and therefore making it harder for senior management to determine whether the strategy was working or not.

In the last couple of decades, a few strategic operational planning methods have been developed. Most of them were derived from two of the more prominent methods: Hoshin Kanri planning (created from the Lean initiatives) and BSC (originally developed by Robert Kaplan and David Norton and published by the *Harvard Business*

Review). These methods provide operations managers with the tools to (1) better define the objectives of a company's strategic plan down to the department level and (2) better linking to departmental planning back to the original strategic objectives of senior management.

In this chapter, I provide an overview of these two strategic planning tools from a manufacturing operations point of view. I also share insight into how a manufacturing company can use the MOM activity model from the ISA-95 standards as a template for strategic planning and engage the CoE as part of that strategic planning.

Whether senior managers recognize Hoshin Kanri or BSC as their planning tool, understanding why these tools are of value and the mechanics of these methods can help broaden a manager's understanding of strategic planning in general. A general understanding of these methods will provide a knowledge base for further explanations on using ISA-95 and a CoE in strategic planning.

My first statement in these overviews is that I am not trying to provide a detailed explanation of Hoshin Kanri or BSC. Many good books, management papers, and training courses are available for that purpose. I am providing what I hope is a sufficient explanation of these strategy management tools to provide context for the remainder of the chapter on how the ISA-95 MOM model and a CoE for manufacturing operations can be beneficial during the strategic planning and execution process.

Hoshin Kanri Planning

In Hoshin Kanri planning, the purpose of the planning cycle is to define a means of identifying critical business issues and/or objectives that must be resolved for company success and then to define a set of detailed activities down to the department manager level that will specifically address these issues or objectives. Objectives/issues may have been determined as a result of a gap in capability from S&OP, for example, or more likely as a result of performance concerns from the operations perspective. In the Hoshin Kanri management mindset, these objectives/issues are well documented, including the symptoms of the issues and the business impact. With a clear definition of the concern, each issue is then discussed in meetings to ensure consensus on the critical nature of the issue and an understanding of how the issue is being tracked (measurements of performance and what changes in the performance are required).

As shown in Figure 7-1, each objective/issue is recorded in an annual planning table that also defines, in general terms, the direction that senior management wants to take to either resolve the concern or prioritize a different concern. The annual plan is then

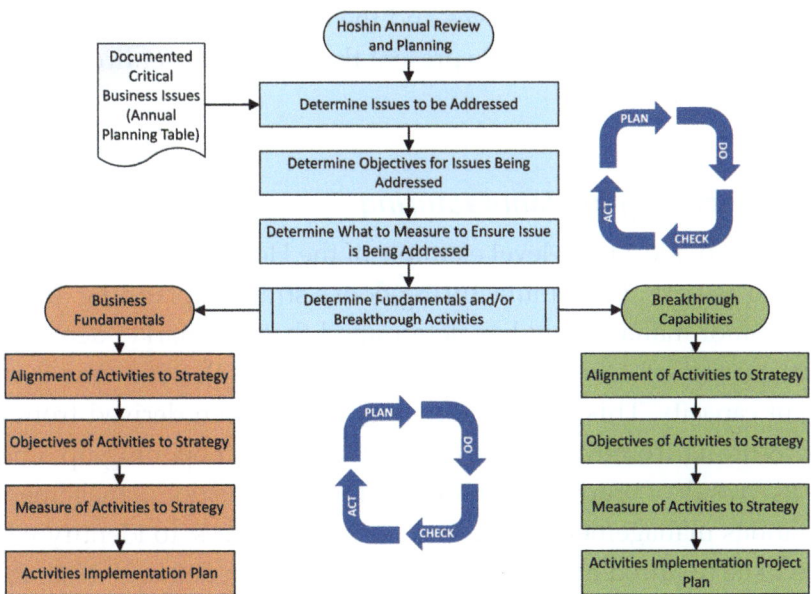

Figure 7-1. Overview of Hoshin Kanri planning.

reviewed and clarified with lower-level management (a process known as *catchball*). With the annual planning table established, the contents are then passed down recursively to lower levels of management as either part of the business fundamentals or the breakthrough capabilities. They are to discuss the direction required to achieve each of the objectives (creating a list of specific lower-level objectives) and to link the strategic activities of the lower department heads to each of the objectives in the annual planning table.

The result is that the objectives for each department are directly linked to those of their senior managers. In a company where the Hoshin Kanri process is mature, it is possible to link objectives from senior management to objectives for each employee, resulting from a deep knowledge of the planning process, even at the employee level.

With each department's strategic objectives linked to higher-level objectives, the department heads develop plans for initiatives or projects to achieve each of their objectives. After the implementation plans have been developed and approved (including costs, time required, and resources), it is time to begin executing the activities and monitoring achievements. Any issues found are documented and reported to higher management on an ongoing basis until the next annual review process starts. Keep in mind that this is a *highly* summarized explanation of the Hoshin Kanri planning process. A key aspect of the entire planning process is the recognition of continued monitoring of the achievement of lower-level objectives, the impact of those lower-level achievements (or lack of achievements) on the higher levels, and a constant evaluation of the soundness of the higher-level objectives and the original strategy. The concept of the Deming cycle of Plan-Do-Check-Act is present throughout the entire process between all levels of management.

Hoshin Kanri Planning in Manufacturing

In this section, I provide a high-level example of the Hoshin Kanri process as it might be implemented as part of a manufacturing floor's strategic planning. In the example in Figure 7-2, senior management has determined that the market demand for a particular product (product A) will be 20% greater for the following year than the company's current capacity. This determination may have been derived from marketing estimations resulting from the product's popularity, from opening up a new market segment, or as a result of production underperformance. Senior management will then look to operations management, using the catchball process, to identify what change in capacity will be required to support the 20% increase in demand. Note that without detailed knowledge of current line capacities, it is impossible for senior management to identify the actual capacity requirements. In addition, most capacity statements that senior management uses are estimated as a dollar amount (e.g., a capacity of $1.3M) based on some standard product mix.

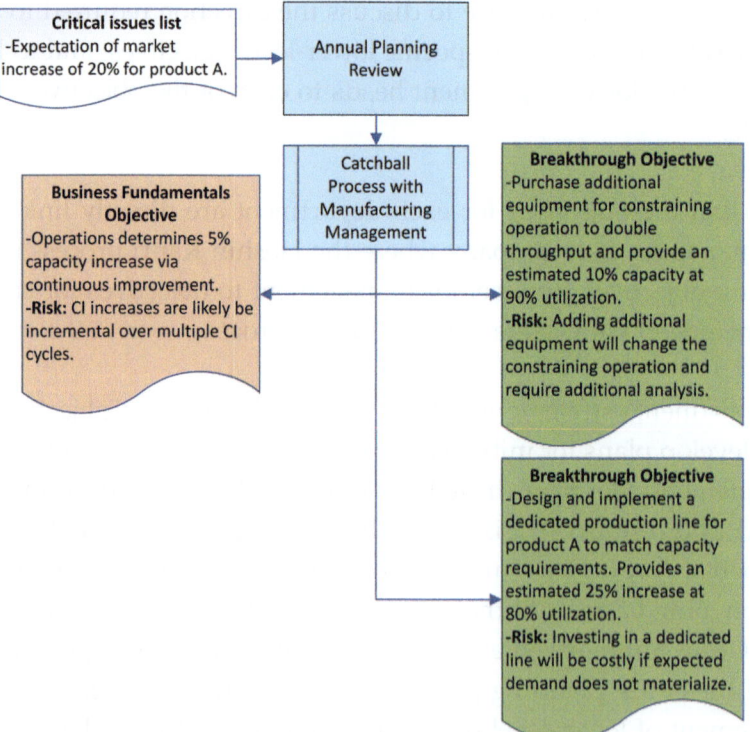

Figure 7-2. An example Hoshin Kanri planning process with a single objective.

In this case, one of the first things that Manufacturing Operations ("Operations") will want to know is whether there is a change in the product mix. In the strategic planning sessions, decisions may be made to phase out products or change the mix of product families resulting from a change in customer order normalization. With a detailed understanding of the product mix that has been demanded, there is a better understanding of the needed capacity for a particular product. In some cases, the increase in demand for one product may result in a drop in demand for a different product, which creates the opportunity to transfer some of the available capacity to product A (provided there is also a comparable demand for resources for each of the products). When there is an understanding of the product mix, Operations can look at what changes will be required to support the expected increase in demand for product A.

While looking at the current process for the product (in Lean terms, they would be looking at the full value mapping, not just the process), Operations can consider implementing some continuous improvement initiatives to help with added capacity (keep in mind that only by improving the constraining operation can you improve capacity). The feedback from this line of analysis is that a 5% increase in capacity can be expected because of better process handling and continuous improvement under the business fundamentals category. However, it is also noted that it will take a few continuous improvement cycles to obtain the entire increase in capacity. The feedback from the line analysis also makes the statement that any further required increase will need some aspect of capital investment and "breakthrough" initiatives.

At the same time, analysis will be performed to identify changes that can be made as a result of capital investment (usually by adding resources, such as an additional test system, to the production line) to dramatically increase the throughput of the constraint operation. This analysis identifies breakthrough changes to significantly increase capacity, how much change in capacity can be expected, and how much the change will cost. The risk that comes from this line of analysis is that when there is a breakthrough change in the constraint operation, there is a high likelihood that a new constraint will appear in a different operation (resulting from the changes in the original constraint operation), and the full capacity needed may not be achieved by the single capital investment initiative.

Depending on the volume of product that is going to be demanded, Operations may also look at an additional aspect of breakthrough capacity. If the demand volume is high enough, there may be justification to design a dedicated line for product A with the required capacity. The risk with the dedicated line is that a new production line takes time to implement (and may affect when that capacity can be available), and if there is some uncertainty in the expected demand (the original 20% increase may be uncertain), the new production line may not be fully utilized, reducing the return on investment (ROI) on the cost of the production line.

Some Operations managers make a mistake in assuming that senior management is interested in a single recommendation that only achieves the end goal of the 20% capacity increase. However, by using the catchball process properly (open communication between levels of management and other departments), questions raised by Operations managers should drive further analysis of the confidence of the original increase in demand (not to mention the possibility of sudden changes in the market). In the Figure 7-2 example, the Operations management team will be responsible for providing senior managers with feedback on two options to deliver the capacity increase, the cost of each option, and the risks of the options. In sending the analysis back to the senior managers, catchball should result in questions about the confidence in the change in market demand with consideration for contingencies based on how quickly the demand is to be realized and the confidence level in the final demand requirements. This process of passing information back and forth should be normal for any good strategic planning. Implementing and using the catchball process formalizes the communication and makes it more transparent, which in turn helps eliminate miscommunication.

Recognizing the significant investment in creating a new line (the second option for breakthrough objectives), there will probably be some reluctance to accept this option unless there is considerable confidence in the expected demand materializing and being sustained. In this case, the direction from senior management will be to maximize the business fundamentals objectives (the estimated 5%) and implement the purchase of additional equipment to update the constraining operation and highlight the new constraining operation as soon as possible. In preparation for the need for a new line, Operations could be instructed to size and design the line without making purchases. Once the demand shows an indication of being sustained, much of the preparation work will already be completed.

Because the two primary operations objectives are to achieve a 5% increase in capacity via continuous improvement and update the equipment resources to achieve a 10% increase (for a total increase of 15%), these objectives will be approved and incorporated into the Hoshin Kanri implementation plan and included in the periodic reviews to ensure the scheduled delivery on the plan.

Planning for Business Fundamentals

In planning for achieving the objectives of the business fundamentals track, the Operations team will look at prioritizing activities to improve the current process.

Planning will also include activities such as updating the value stream mapping for the product, analyzing the capability and fallout, investigating non–value-add activities such as wait times, and using reports such as Pareto charts (discussed in Chapter 3) in selecting specific nonconformances and inefficiencies to derive greater detail for the planning of further initiatives. Assuming that wait times have been minimized (one of the first issues resolved in Lean initiatives), after the specific nonconformances have been selected using Pareto analysis, machine learning can provide insight into the process parameters that have the most significant impact on the fallout.

This kind of planning requires a great deal of focus and consistency in the administrative and analysis processes for continuous improvement, and the ability to plan for these kinds of initiatives is a significant reason for developing a skilled CoE. Although analyzing the product's value stream map is the responsibility of the manufacturing engineer assigned to the product, the CoE's guidance makes this kind of initiative planning possible. In Lean analysis, it is acceptable to make improvements on a wide variety of issues that can improve the cost or cycle time of any operation in the process. The important aspect of determining the issues causing concern about the business fundamental's objective is to ensure that the concerns investigated as part of continuous improvement are directly related to improving the *throughput* of the process. *Any continuous improvement activities that do not improve the process's throughput should be a lower priority.* This is an important facet of Hoshin Kanri. Directly relating the activity of business fundamentals to strategic objectives provides guidance to operations activities that will have a direct effect on the strategic objective. If the strategic objective is to lower the cost of producing product A, then any continuous improvement activity that lowers the overall cost of producing the product will be of value. With the planning objective being related to capacity and throughput, reducing the cycle time of the constraint is the better direction to take in initiative planning.

Planning for Breakthrough Objectives

In planning for achieving the breakthrough initiative, the Operations team will look at the same value stream mapping as that used in the business fundamentals objective (for simplicity, again assuming wait times have been minimized), but this time they will focus on the constraint operation (example in Figure 7-3). In this section, I provide an example to explain the concerns Operations is facing with reaching the breakthrough objective.

Figure 7-3 is a simple example of a *one-piece flow* process in which the cycle times of each operation are similar (very little difference in the time differentials). However, note that even in this process, there is still a constraint operation (Operation 2 @ 50 s/

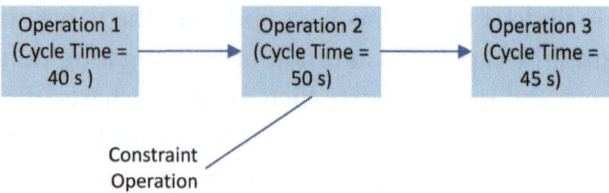

Figure 7-3. A sample process with a single constraint operation.

unit) that defines the throughput of the entire line (usually defined as a result of the planned takt time of the overall process). Assume that the current line is operating three 40-hour shifts at 90% capacity (Operations understands that maintaining 100% utilization is not sustainable and would not be used for planning purposes), which would deliver a little less than 7800 units per week (3600 s per h/50 s per unit = 72 units per h, 72 × 40 × 3 = 8640 units per week [total capacity], 8640 × 0.9 [90% capacity utilization] = 7776 units per week). To deliver on the breakthrough objective, the change in capacity must achieve a throughput of a little less than 8560 units per week. The option chosen for achieving that throughput is to add a second resource at Operation 2, which would enable that constraint operation to produce at effectively half the cycle time of approximately 25 s, theoretically doubling the throughput. Or does it?

The change in the resources working on Operation 2 effectively reduces the joint cycle time of that operation to about 25 s (as shown in Figure 7-4). However, as was suspected during the initial planning (indicated by the risk statement), Operation 3 now becomes the new constraint at 45 s/unit, making the new line throughput at 90% utilization a little more than 8600 units per week. Looking at throughput, you have achieved the required 10% increase in production; however, this assumes a 100% FPY. We still must consider that the FPY is not likely to be 100%. If the business fundamentals initiative does not achieve at least a 99% FPY, throughput will drop below 8550 (8640 units × 0.99 (99% first-pass yield) = 8553 total units), putting the breakthrough objective at risk of still not being achieved. Achieving a 99% FPY will now place additional concern on the business fundamentals side.

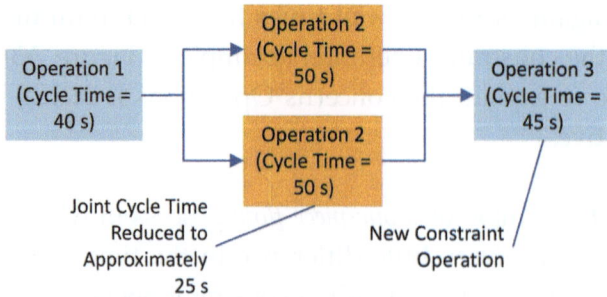

Figure 7-4. A sample process with an additional resource.

One issue this example highlights is the dependency that strategic objectives may have on each other and the importance of having accurate numbers and context during the planning process. In this example, only one strategic objective was defined. In any normal strategic planning session, three or four objectives (the best practice in strategic planning is not to establish too many objectives at that level) may be defined at the senior management level, which can make the planning and monitoring processes much more complex.

Balanced Scorecard Planning

The balanced scorecard (BSC) planning process is similar in many ways to Hoshin Kanri. In fact, many believe that combining the two can produce a much more rounded strategic plan. Although there may be some truth to that line of thought, that combination also makes for a much more complex strategic planning process, and it should probably be considered only by companies that already have mature strategic planning processes established.

In BSC planning, the executive managers establish a mission and vision and then define a strategy to provide direction on objectives for achieving the mission and adhering to the vision for long-term growth. One of the main aspects of BSC (as shown in Figure 7-5) considers the objectives not only from a financial context (which is typical) but also from the context of (mainly) three other perspectives: customer satisfaction, process improvement, and internal learning and growth. These three are also implicitly considered as part of the Hoshin Kanri method, with process management and customer satisfaction being explicit components of Lean and continuous improvement by applying the Deming cycle at all levels of planning. In BSC, senior managers should consider strategic objectives to be used in support of the company's mission (their definition as

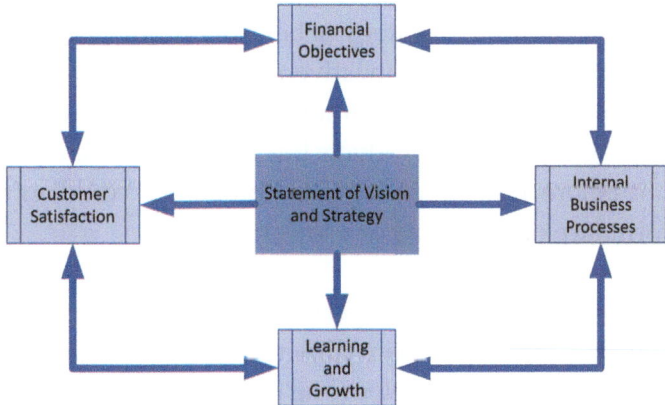

Figure 7-5. Planning objectives of a BSC.

to why the company exists in the first place) and the company's vision (how the mission will be accomplished over the long term) from the perspective of all four contexts. As a result, if senior management sets an objective for increasing sales revenue, for example, the objective will also require consideration (and setting additional objectives) in the context of customer satisfaction (e.g., improve repeat customers by a specified amount), which internal processes the company must develop (or improve), and which new skills are required to achieve that growth. The mindset, in this case, is that unless objectives were set (and achieved) in all four contexts, the original strategic financial plan is little more than a *wish list* with no guidance as to how it will be achieved.

In Hoshin Kanri, the tools (templates, tables, etc.) for linking objectives to lower-level activities are well defined, but in BSC, the tools are not as specific. Using "strategy maps" helps BSC practitioners to visualize the objectives, but a specific format for defining links to objectives is not defined.

BSC in Manufacturing

Similar to the section on "Hoshin Kanri Planning in Manufacturing," in this section, I provide a simple example of using BSC as it pertains to the planning for a manufacturing floor.

In BSC, there is a lot of emphasis on deriving the scorecard from objectives set by the company's strategy. (Scorecards can be derived just from predefined objectives without links to the strategy, but this is not recommended because objectives defined without a strategy lack the focus needed to tie the objectives together.) So, I define a scenario and a change in company strategy (for a hypothetical company, of course) and provide an example of the BSC that might come out of a manufacturing company.

BSC Company Scenario: Example

Company X has always followed a strategy of being a product leader and therefore has been a market leader in providing innovative products that are difficult to replicate. This strategy gave company X the lead in market penetration long enough to not only recoup the engineering costs of the products but also to provide a period of good ROI for company profit before innovation-lagging companies could introduce products with similar performance or capability at a lower cost.

In recent years, however, the company's period of leading innovation for its products has been getting shorter, making it more difficult to maintain that level of ROI. After some research, company executives determine that changes in technology have made it easier for competitive companies to introduce comparable product features, making it more difficult for the company to maintain the strategy of strictly relying on product leadership. Company X's executives determine that the strategy of product leadership must be matched with improvements in operational efficiency to (1) introduce products to market faster, giving competitors less of a heads-up on new products; and (2) introduce products at a lower initial price point, making it harder for competitors to introduce comparable or better product value.

The chief operations officer (COO) of Company X has chosen to implement Lean as the primary management methodology to improve operational efficiency and to do so in two phases. The first phase will apply to the manufacturing floor because it has a greater awareness of process management in general and should be able to establish a formal program quickly. After phase 1 is well underway, the second phase will expand Lean to the remainder of the company's operations. (*I recognize that company X could also introduce Lean for Design to help with new product introduction, but this is a book about manufacturing operations.*)

As stated earlier, the strategy map has become a significant part of the BSC implementation. It is used extensively to cascade objectives from higher-level management down to line managers and even operational staff. Figure 7-6 shows the initial strategy map for the specific objective of improving operational efficiency. In this case, the objective has been expanded into three of the four primary categories. (Although a customer focus objective also could have been stated, I'm keeping the focus of this section on the operational objectives.)

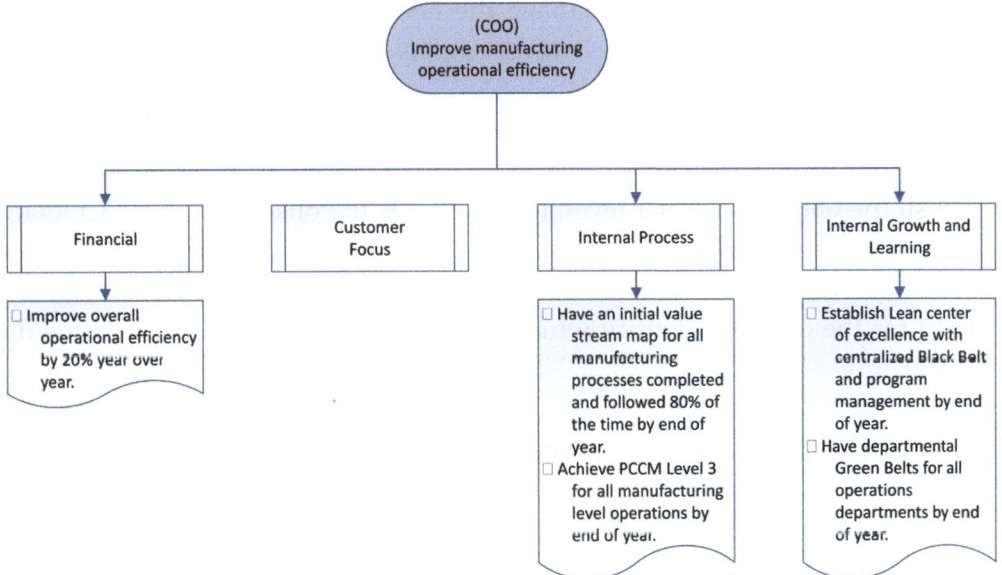

Figure 7-6. Initial strategy map (operational efficiency).

Now that these objectives have been defined, they are passed down to lower-level management so they can establish their own objectives appropriate for their level of management and aligned to the objectives of Figure 7-6. To simplify the cascading of operations objectives, I will focus on the manufacturing branch (darker boxes in Figure 7-7).

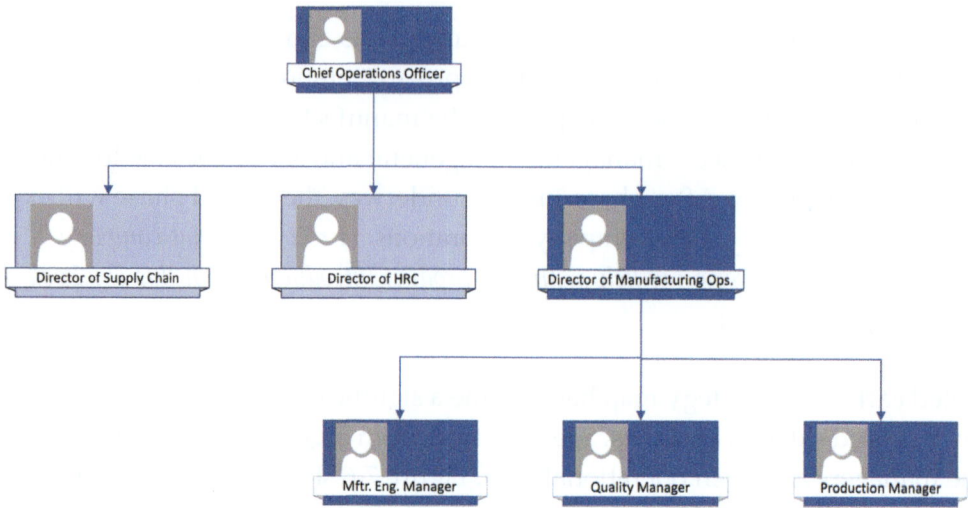

Figure 7-7. Company X operations organizational chart.

The initiative for Lean is determined to be extremely important, and the COO is providing direct oversight. For the objective of internal growth and learning, the COO will directly establish and manage the Lean program. This includes establishing and managing the Black Belt–level employees and managing the initiatives under the Lean program with department-level employees being certified as Green Belts. As stated earlier, the objectives from the COO level are cascaded to lower-level management. The director of Manufacturing Operations works with the COO to establish the objectives at her level. In Figure 7-8, the objectives at the COO level have been defined with a unique identifier (in this case, A-level objectives are from the executive level; other letters are used for lower levels of management), and the objectives for the director identify which of the executive objectives are linked (or cascaded from).

It would be difficult for the management staff to support the certified Lean staff in their areas without understanding Lean, so the director of Manufacturing Operations has an objective for the certified staff as well as introductory training for the management staff in her department.

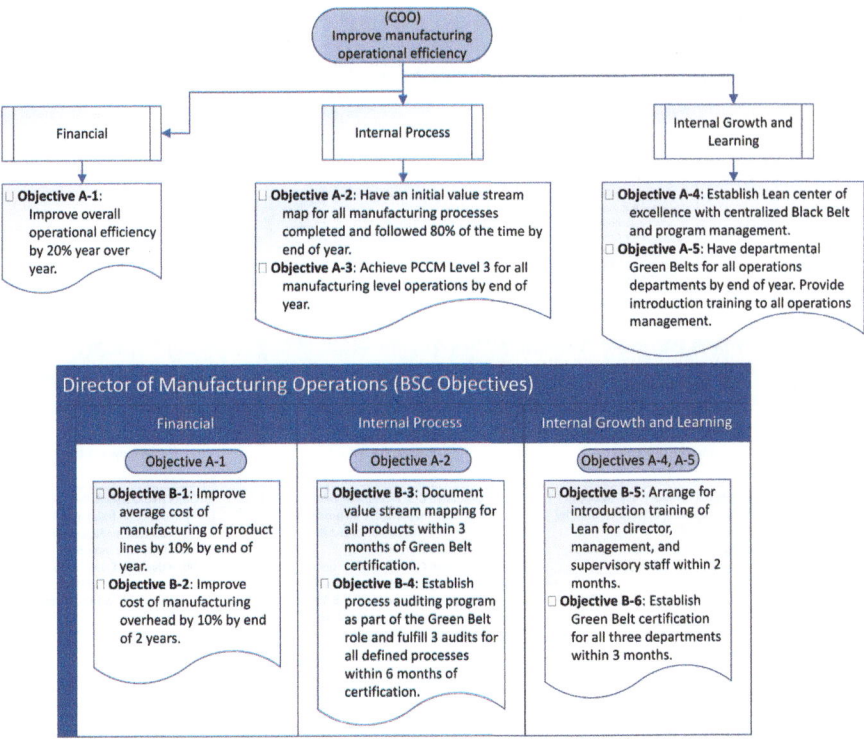

Figure 7-8. Cascading objectives in the COO strategy map for company X.

Cascading BSC Objectives

When the director of Manufacturing Operations' objectives are cascaded to the production manager (Figure 7-9) and his organization, the manager will have an internal growth and learning objective to get the introductory training for himself and each of his supervisors, as shown in Figure 7-10, and will also train one member of his staff to be a Lean Green Belt. The internal processes for all lines on shifts 1 and 2 will need a value stream map. The timeline will be passed down from the director's Objective B-3, as will the processes for test and repair, and all these processes must engage in the process auditing defined by the director's Objective B-4. Each of the production manager's objectives will also be uniquely identified.

As the production lines are audited and the team (the Green Belts from all three groups) works on removing "non–value-added" steps from the processes, the director's Objective B-1 is not expected to be too difficult to achieve. However, the work-in-process inventory for nonconforming products (products needing rework or repair) has long been a concern (it frequently is with many companies). The production manager (and the test and repair supervisor) commits to reducing this inventory by 30% by the end of the year by implementing a faster process for repair as a link to Objective B-1.

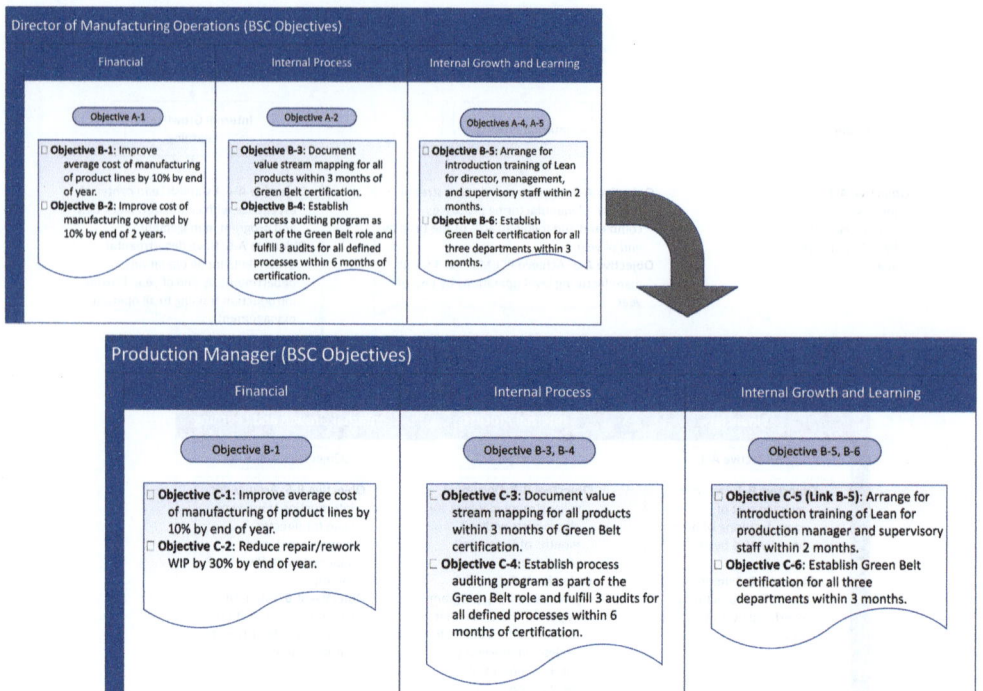

Figure 7-9. Cascaded BSC from the director of Manufacturing Operations to the production manager.

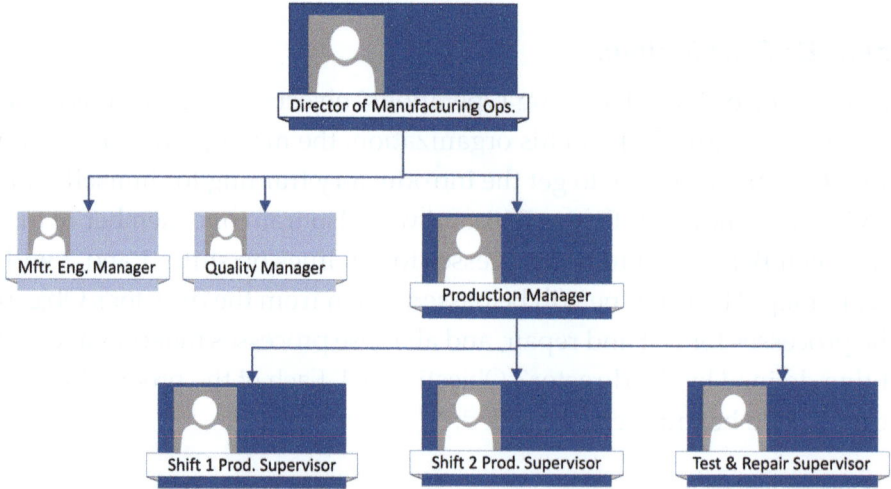

Figure 7-10. Production-level organization chart.

There are two key aspects of BSC presented in this example: (1) the strategy defined includes more than just financial objectives, which forces the owner of the objectives to think about how financial objectives are going to be met and not just "meeting the numbers"; (2) the strategy ensures that the objectives of lower managerial levels are linked directly to the objectives of the more senior managers (and all linked to the strategy of the executive level).

Strategic Planning and the CoE

One of the factors that creates a problem for strategic planning (regardless of the methodology) is that as objectives are passed down to departmental managers, there is a strong tendency for the cascading lower objectives to focus on the scope of the departments themselves, which is understandable because it is the scope for which the manager has control. However, as in the preceding example involving the production manager and the repair and rework supervisor, taking a narrow scope, such as that of reducing the time to repair, can be detrimental to other objectives.

Cascading Strategic Objectives

Although it was not presented in the previous example scenario as a link to the manufacturing director's objective of lowering the cost of manufacturing, the quality manager would likely have an objective of increasing the FPY because increasing FPY also reduces the number of production units that require repair or rework (eliminating the extra cost of the hidden factory, introduced in Chapter 2). This kind of objective requires a lot of analysis and a lot of data for that analysis. If, in the process of reducing repair times, the repair and rework group drops the data collection on what action was taken during repairs, its objective may be eliminating the data needed for reducing FPY. There is, of course, a point where more data for a particular problem becomes redundant. But without coordination between the quality group and the repair and rework group, the loss of data collection in the repair and rework processes will impact the analysis of increasing FPY and other quality initiatives.

When a company takes the time to develop a CoE with the full scope of manufacturing, there is a stronger tendency to reflect on strategic objectives that are broader in scope and therefore accept initiatives that require greater intergroup collaboration. This is derived from a broader understanding of the analysis processes used in continuous improvement efforts and a greater understanding of the impact that focused objectives can have on initiatives in general.

Revisiting the company X scenario (which could be reflected in both BSC and Hoshin Kanri), it is important (and likely to be much more impactful) to develop strategic initiatives that are focused on the output and throughput of production in general. In initiatives like Lean some common errors are in the localized reduction of non–value-add activities; the repair and rework scenario presented earlier is a typical action taken. As presented in Chapter 3, when there is a common occurrence of an issue needing repair, before any action is taken to improve the repair time, it is first important to characterize the problem (e.g., how frequent, repair cost [process and replacement material], and mode of failure). After the problem has been characterized, a plan must be defined for how to handle the problem (mitigation or elimination, or at least temporary mitigation until further analysis can be performed). Only after these steps have been performed can a decision be made as to whether to continue data collection on the problem. These decisions should be made at the CoE level; even using an informal group of diverse knowledge in operations is better than the single-department decision scenario.

Defining the Executive-Level Objectives

One of the advantages of developing a CoE is that the manufacturing floor will gain a much clearer understanding of the performance capabilities. It also becomes easier to present this understanding to the executive level. In Chapter 5, I referenced a popular statement that was used years ago: "Visibility from the shop floor to the top floor." When this statement was originally used, it was about using the IT systems to make data visible through all IT systems (system integration) and maintaining the integrity of that data (the common interpretation of the data in each system). Putting aside the data integrity issue that I discussed in Chapter 5, there were other issues that created considerable problems back then (and still do). With data being available to other systems, there exists the issue discussed in Chapter 2, where data objects can have the same name but are interpreted differently (like the "operation" example in Chapter 2). However, when different departments investigate issues using this data, there is a high likelihood of the information (or reports) derived from this data being interpreted differently as well. As department heads share information and discuss reports with their senior management, these differences in interpretation can be complicated even further. Consequently, at the senior management level, the same data might be reported slightly differently up the departmental communication chains, leading to different perspectives on how to handle concerns strategically.

In an effort to reduce (or even prevent) the issues of data interpretation, the CoE can (and should) take ownership of defining how reports are to be interpreted. There will be times when tactical reporting must be implemented and interpreted locally (and outside the scope of the CoE); however, any reports that are used for strategic objectives should be defined by the CoE. As the CoE gains knowledge of the processes in manufacturing (or in business) and as its role is expanded, it is likely that the CoE will become the primary group for performing the high-level analysis used in strategic planning.

Using the CoE and ISA-95 in Strategic Planning

When engaged in the planning cycle of any initiative, there is considerable visibility of the processes that are needed for manufacturing (all the steps to "get it done"). The aspect of planning that frequently catches managers (and sometimes even more senior executives) by surprise is the extent of planning necessary to keep manufacturing running smoothly (i.e., all the support activity). It is one thing to plan the process of making a product, but there is also the need for plans on continuous improvement cycles (the process will not be perfect the first few times through), the availability of resources for equipment maintenance, and training the production personnel and the support staff. To ensure that all areas of concern are covered, it is best to have a template or checklist to validate against. Although the template will likely be unique to each company, using the ISA-95 activity model as the initial template can go a long way to ensure the complete scope of planning and a more realistic plan for achieving strategic objectives.

Another area of complexity that makes planning for objectives difficult is determining the "capability" of the manufacturing processes. As mentioned in the previous paragraph, the initial processes for a new product, new line, or a new plant will not be optimal from the start. All processes will go through stages of improvement over time. Having a historic understanding of the ramp-up of previous initiatives will provide some input for any current objectives. If the CoE has been properly managed, this historic information will be available. As a result, the executives at the strategic planning level will find it beneficial to engage the CoE in strategic planning. In addition, if the company has improved its process management to Level 4 or 5 of the process capability maturity model (PCMM) framework, not only will the "ramp-up" information be available but the CoE's knowledge of implementing initiatives will improve the implementation of objectives so that the manufacturing capability starts at a higher level and the time needed to optimize that capability will be reduced.

Summary

In this chapter, I briefly explained the activity that goes into strategic planning and provided a couple of examples of how that planning may play out for a manufacturing floor.

When thinking about strategic planning, many people imagine the top executives going off to a secluded place and scheming about how the company is going to take over the world. (*OK, maybe it's not quite that bad.*) However, many companies still view strategic planning as an "executives-only" activity that is not open to the rest of the company until the planning is complete. The mechanics of strategic planning methods like Hoshin Kanri and BSC illustrate that "good planning" requires considerable (and realistic) knowledge of a company's current capabilities from a wide range of perspectives. Using strategic planning methods like Hoshin Kanri or BSC helps to ensure that the planning becomes more thorough and reliable, and by engaging a CoE in the planning process and using models like ISA-95 as a template, the objectives defined will be much more realistic and achievable.

Many of the concepts discussed in this chapter can benefit from a well-developed CoE, but it is well worth it to engage a CoE as early as possible when developing the objectives, to take advantage of tools like the planning methods introduced in this chapter, and to use models like ISA-95 and PCMM. Establishing a CoE ensures the development of a knowledgebase that can be used for strategic planning. Engaging the CoE early in strategic planning helps to ensure that there is common understanding of the strategic objectives across departmental lines and a broad understanding of the action needed to support the objectives.

8

Connecting
the Dots

In previous chapters of this book, I discussed what a CoE is, how a CoE can be structured as it relates to different industries (particularly a manufacturing floor), and different methods of introducing a CoE into a company (e.g., developing a CoE as part of other initiatives like Lean or ISO-9000, or as an MES implementation). I have also presented several topics with the intention of providing insight into the core of each of these topics as it pertains to manufacturing and MOM.

In this chapter, I expand on the subjects covered earlier in this book and describe industry best practices that I have found useful and *why* they are considered best practices.

Is It Best Practice or Common Practice?

The first issue I want to clarify is the difference between *best practice* and *common practice*. I have attended far too many presentations from consulting companies describing ideas they refer to as *industry best practices*, only to find that they were simply common practices within that industry. Usually, the so-called best practice was actually a common practice used to take a shortcut in process or product traceability management. The shortcut would tend to summarize information (and lose details of that information) but because everyone used the shortcut, it was considered acceptable to the industry.

An example of a *common practice* in discrete manufacturing is using "phantom" bills of material (BOMs). In this scenario, a subassembly must be completed before it is assembled into the main product. However, the subassembly only contains a few components, so the assembler who installs the subassembly is usually tasked with making the subassembly first at their workstation, one at a time, before assembling the product. At the planning level, the material and processes for making the subassembly would be included in the activity that accounts for installing the subassembly, i.e.: the planning BOM at the enterprise resource planning—ERP level would include the consumption of material, labor, and any other minor resource used to make the subassembly in the ERP-level operation and to install the subassembly. In this scenario, all aspects of making the subassembly are considered a "phantom" of the higher-level assembly. The reason for using phantom BOMs is to simplify planning a production run. One of the issues with phantom BOMs is the potential of losing visibility of the subassembly quality before it is assembled into the product. In some cases, the process of putting together the subassembly first is so reliable that it is not worth worrying about, but this is not always the situation.

In this case, *best practice* is achieved when interfacing MES and ERP systems. At the ERP level, the material and labor are included at the higher product level but are identified as "phantom material." When the planning BOM is sent to the MES (or, in some cases, validated against a manufacturing BOM), the *phantom level* (material and process) is separated into a specific process at the MES level.

> Change (how the floor changes and the rate of that change) is unique to each plant floor and depends on how willing the manufacturing floor culture is to accept change.

Understanding the difference between best practice and common practice can be critical to quality management and is essential at the CoE level. It is also important to ensure that any consulting firms hired understand this difference. If a team member (whether part of a consulting team or internal for any initiative) starts talking about a "best practice," it is worth challenging them to explain *why* it is a best practice; if they are unable to explain the *why*, they may be describing a common practice instead.

The Maturing Manufacturing Floor: Changes Post–MES/CoE Implementation

In a few chapters, I introduced directions for change "as the company matures" after implementing an MES and/or a CoE, but what are the changes that a company experiences?

An important aspect of change a manufacturing floor will experience is that change (how the floor changes and the rate of that change) is unique to each plant floor and depends on how willing the manufacturing floor culture is to accept change. In the years that I have been working as a consultant in the MES/MOM sector, I have met company employees who found that change is exciting and helps to make the job more interesting. I have also met employees who resisted change "tooth and nail" (as the saying goes). The number-one rule that must be accepted is that change is inevitable. For people who are associated with a changing environment in one way or another, I offer the words of a colleague of mine: "Change is going to happen, so you must either lead, follow, or get out of the way." With any initiative (e.g.: Lean, ISO-9000, MES, or CoE), there will be changes. Some of the changes will be simple process changes and easy to adopt. A company's ability to do well with these initiatives is not related to which MES is implemented or whether it is implementing Lean or Theory of Constraints; the companies that succeed in implementing an MES (or any other initiative) recognize and help implement a change in *culture*.

When a manufacturing floor implements formalizing a process (by implementing an MES, CoE, or any process mapping methodology), the first hurdle the initiative's team will face is the recognition of how chaotic the manufacturing floor is at first. Without a defined process (and the constant verification of that process) using formal process management methods, the production operators on the manufacturing floor will self-optimize. They *will* develop through repetition a process of doing the work that "works best for them at that operation." In reality, it is not an *optimized process* but a *familiarized process*, and each production operator will *familiarize* a process in their own way. Therefore, when the process documentation begins, there should be no surprise about how much variation a process has. It is important to capture (document) these variations and why they are performed.

While defining and implementing the initial process, the key aspect is to make the process repeatable and to gain an understanding of that process's characteristics (capacity, fallout, cycle times, etc.), as discussed in Chapter 3. As part of the cultural change, the production operators must learn to follow a process as diligently as possible and not fall back on

> It is only when a process is repeatable *and* reproducible that it is considered stable.

their "familiarized" process. This is an area in which an MES can help by ensuring that steps are taken in a specific sequence. Because this is likely to be the first time that detailed process characteristics are measured, the next hurdle for process formalization is accepting the characteristics the MES reports. There will be attempts to discredit the characteristics that have been recorded, and in some cases, the method of data collection and/or the data interpretation may actually be wrong. The MES does not record and report on the correct data; *it reports on the data it is told to report on*. If there are errors in the data collection and interpretation, then there must be an effort to improve the data collection or the interpretation—*do not get rid of the system or process.*

It is only when a process is repeatable that the process characteristics are recorded and interpreted accurately, and *the culture of the production floor has confidence in the information presented* that real improvement can begin.

Another concern that can create confusion is the process aspect of being "reproducible." What is the difference between repeatability and reproducibility? In a single production run in discrete manufacturing, a sequence of steps is performed for each production unit, and this sequence is (hopefully) repeated for each unit in *that* production run. This is repeatability. The next day (or week or other time frame), the production floor will set up from scratch and perform another production run of that same product. On each of the following production runs, can the production floor *reproduce* the results of the first production run? It is only when a process is (more or less) repeatable *and* reproducible (remember the concept of variation explained in Chapter 3 in the section "Using Pareto Charts"?) that it is considered stable. Most of the activity in the early stages of process management (the activity of getting from Level 1 to Level 3 of the process capability maturity model—PCMM) will be directed at getting the processes to achieve this stability. Process stability also must be achieved in all four pillars of the MOM activity model presented in Chapter 2. Defining the process management activities highlighted in the ISA-95 activity model (activities related to performance analysis) helps ensure the management of processes in each of the four pillars.

When a production floor has gained a level of stability, its culture should also have changed to recognize the importance of that stability and of being able to *measure* process characteristics. After a production floor has achieved this level of maturity in cultural change, initiatives can be taken to improve specific characteristics of the process (moving from Level 3 of PCMM to Level 5). *The greatest hurdle in achieving high performance in process and operational cost efficiency is changing the culture of the production floor.*

Management of Continuous Improvement: Reprise

You have engaged in the activity needed to stabilize your processes (both repeatable and reproducible), and you have *relatively* repeatable measurements of your processes. Now what?

Taking guidance from Chapter 3, the process management activity can change from stabilizing processes to improving specific characteristics of the process. The keywords here are *specific characteristics*. On many occasions, I have observed Lean initiatives, in particular (although I've seen it in many programs of continuous improvement—CI), having the production floor try to improve everything at the same time. One concerning aspect of improving processes is not what to improve but determining the effects of each improvement initiative. As changes are made to the process to improve a specific characteristic, interactions that will affect other process characteristics are likely to occur. These interactions of process change must be minimized if real improvement is to be made, and as presented in Chapter 7, the initiatives taken to improve a process should be linked back to a strategic initiative related to either cost, capacity, or product quality improvement. In any CI initiative, "what is being improved" must relate to a *management decision* to improve a specific aspect of operational or product performance.

In Chapter 3, I presented several concepts in managing CI, such as the four stages of improvement (detection, correction, prediction, and prevention), standardization, and determining which issues to improve first (using Pareto charts). I also discussed one of the benefits of standardization: the reduction of noise (reduced variation), which eases the efforts of discovering issues. After an issue has been uncovered, the next problem in CI is identifying the root cause. One of the inhibitors of identifying a root cause is access to data.

Master Data Configuration

Two concerns that influence the ability to improve the process or product are access to data and isolating that data to specific problems. An MES can help in this regard, but it still depends on the master data configuration. If data is not being collected, it, of course, cannot be analyzed. As part of the analysis, the data must be "sliced and diced" to focus on the correct parameters that describe the problem. As initially presented in Chapter 6 in the section "MES in Discrete Manufacturing," care must be taken when configuring data objects in the system. The way a physical production floor is defined in the virtual world of information technology (IT) systems will have a significant impact on how distinctly data can be extracted.

Figure 8-1 is an example of the configuration of a production line with each operation in the process defined as a workstation on the production line. In this example, there is no capability to distinguish data that is specific to the operation performed at the workstation from the data describing the workstation itself.

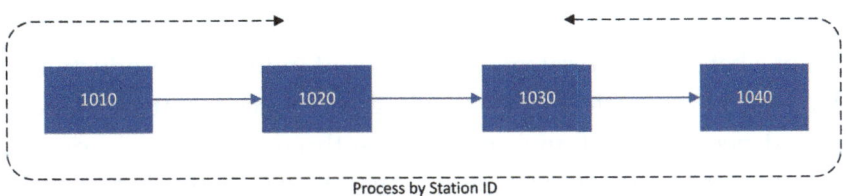

Figure 8-1. Poor configuration of master data.

In Figure 8-2, the operations in the process are configured as a set of data objects separate from the configuration of the workstations on the production line. This enables performance data related to the workstation to be analyzed separately from the performance of the operation. The greater the granularity of the master data within an MES, the more distinct the data analysis can be.

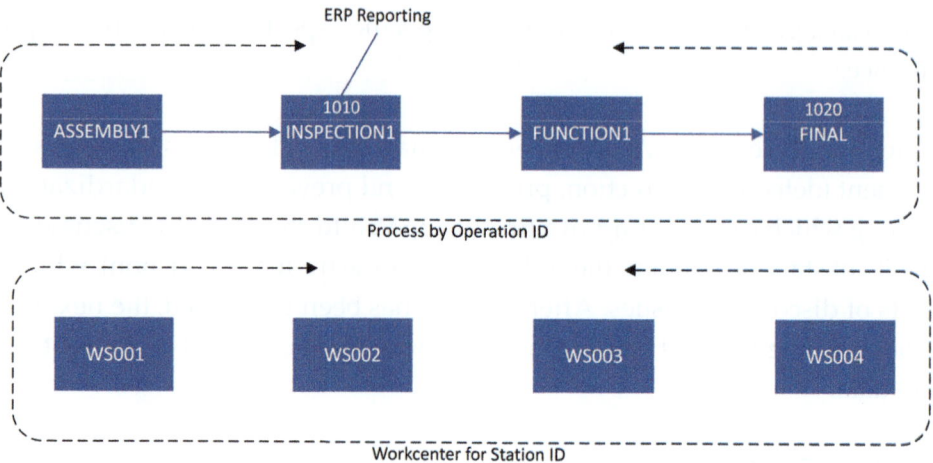

Figure 8-2. Expanded configuration of master data.

Another aspect of master data that can be problematic is production unit identification (not to be confused with product identification). Depending on the manufacturing model (e.g., process or discrete), how each production unit is identified can greatly help or greatly hinder the analysis.

Production Unit Identification

Depending on the product structure, a production unit may be identified by a serial number or a batch (lot) number. If the process allows for a single unit to have its own distinct path through the production process, each unit probably needs to be serialized (a unique identifier for each individual unit). This enables all activity performed on each production unit to be recorded and analyzed. On the other hand, if production activities are performed on larger quantities of production units at one time (all units follow the same path through production), a batch ID (or lot number) is a better choice. However, how the batch ID is defined can also be a problem.

The purpose of a batch ID is to categorize a series of individual production units with very similar process characteristics according to process, material, and resources (equipment and people). So, if any one of these three inputs into a production run changes (or any production input changes the characteristics between production units), theoretically, the best practice would be to change the batch ID as well.

I use the term *theoretically* because, depending on the production environment, it may be difficult to determine and/or track when the batch ID should change. This is another area where *common practice* has taken precedence over *best practice*. In an effort to gain *some* visibility of process changes but also maintain a batch ID system that is easy to manage, companies that use batch IDs have implemented creative identification schemes. Some are as simple as using the shift or week number to help narrow down when the unit was made. During this period, the production floor will record the identifiers of all material consumed and the equipment that was used; some will also go as far as recording the production operators who performed the work. Others will use a combination of week number and production run ID during the week, as an example. A fundamental reason for using an ID of some sort and recording the production parameters used during the production run is to provide some aspect of traceability for warranty issues. By using the batch ID, Quality can narrow the production parameters that may have affected the quality of the product and may therefore be responsible for a warranty claim or product recall. Although these identification schemes are easier to maintain on the production side, they will make the analysis in support of warranty or recall more difficult than an identification scheme based on best practices.

Figure 8-3 presents a simple example of a batch ID management scheme based on best practice. In the example, the production run starts with using Equipment 1 at Operation 1. The batch ID used to start the production run would be 0422000123A, with the prefix "0422" identifying the week number and the year (indicating the week of January 24 to 30 in the year 2022). The sequence 000123 identifies the production run for that product during that week. The suffix "A" identifies that this is the initial sequence of the production run. During the week, the production floor switches (or adds) Equipment 2 (a change in equipment) to the prefix "0422," but now the sequence

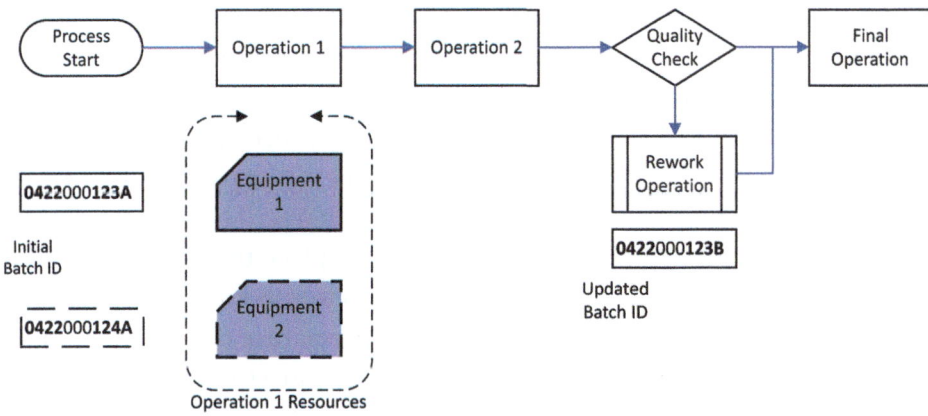

Figure 8-3. Simple batch ID based on best practice.

has been incremented to 000124. Because this is the initial identifier for this production run, the suffix stays the same ("A"). It is important to understand that if product was running on both Equipment 1 and Equipment 2, both batch IDs would be active at the same time.

During the production run of 0422000123A, a quality check determines that some of this production run has a nonconformance and will require rework. The subset of this production run that is directed to the rework operation (a change in process from the original production run) will be relabeled to 0422000123B to indicate that additional activity was performed on a subset of the original production run.

In this example, the production units included in any of the batch IDs will all have the same production activity recorded against them. The concern (expressed earlier) is that managing when to change the production ID can be difficult with a manual tracking system. When using an MES (or other automated function), enabling the system to define not only the initial batch ID but also changes in the batch ID for all process changes will make managing the production ID much easier. Implementing a similar scheme through a central CoE will help ensure consistency across all production plants.

By managing the production identification (serial or batch number) and detailed master data configuration, the Manufacturing and Quality Engineering groups will gain the most detailed virtualization of the production environment and the most significant visibility of interactions on the production floor that may have caused quality concerns.

Symptom versus Cause versus Root Cause

In every quality management function, there is always discussion and analysis to identify the root cause of a problem, but what is meant by the *root cause*? This question is confusing because the answer is actually a matter of perspective. In other words, "it depends." From the customer's perspective, they are unhappy because their cell phone is not working. Their root cause is that the battery will not hold a charge. The solution is to replace the battery (or replace the phone altogether); for them, the problem is solved. However, from the phone vendor's perspective, the battery getting to the point of not holding a charge is premature (the phone is only three months old). The vendor will then follow up with a request to its supplier, who will determine that the battery was part of a bad batch. From the vendor's perspective,

Determining the initial cause at the lowest level of control that is available is how you differentiate between *symptom*, *cause*, and *root cause*.

this is the root cause, and it must check the inventory to ensure that there are no other phones with that batch of batteries. The vendor's problem is solved. The battery manufacturer determined that the machine that filled the batch of batteries with the proper chemicals was not working properly (its root cause), so it fixed the machine. Has the root cause been determined and the problem solved? Well, not yet. Did the machine malfunction as a result of instructions for use, instructions for maintenance, or another issue? These questions must be answered to determine the root cause for the battery manufacturing company.

If you asked the original customer, the *symptom* was that the phone was not holding a charge, and the *cause* was that the battery was failing. From the vendor's perspective, the battery failure after only three months was the symptom, and the bad batch was the cause. The determination of the root cause depends on whose perspective you are viewing. From the phone assembly company's perspective, the root cause may be that the battery supplier is unreliable. To fix this root cause, the assembly company stops using that battery supplier. From the battery manufacturer's perspective, the machine malfunction was caused by an improper maintenance procedure. Fixing this root cause requires a new maintenance procedure.

We have all heard the advice in *root cause analysis* to "ask *why* five times," but is it really that simple? As illustrated in the previous discussion, not really. Determining the root cause requires finding the source of a problem at the lowest level of control available from your perspective (the perspective of the person making the root cause determination). This means that the root cause from the phone assembly company will be different from the root cause from the battery manufacturing company. *Determining the initial cause at the lowest level of control available is how you differentiate between symptom, cause, and root cause.*

Maturing in Process Management

In Chapter 3, I discussed the fundamentals of CI and introduced the PCMM. One of the main CI concepts I discussed is the aspect of being methodical in the steps to improve a process. But what does that really entail? If you look at methodologies such as Lean, Six Sigma, or the Theory of Constraints, they identify the steps for a single cycle of an improvement initiative according to their methodology and use their templates to determine how to analyze an issue and how to ensure that the solution is working. Not often discussed is (1) when you should use a particular methodology (Lean, Six Sigma, or Theory of Constraints) and (2) what overarching process ties initiatives together (and eventually ties initiatives back to strategic requirements). In this section, I go deeper into the activities of process improvement and outline the fundamental steps of any CI program.

Let us revisit the scenario in Chapter 3 with you as the new manufacturing engineer who had to create the process for manufacturing a plastic case. Because you now have some experience in process management, you are assigned the task of leading the formalization of the company's CI program. The big question on your mind is, *where do I start?* Having been attentive to "talk in the industry," you have heard of a process improvement model called the *process capability maturity model* (PCMM), which categorizes maturity in process management into five levels. You recognize it as a model that provides guidance for the activity needed at each level and a means of measuring progress. Partly from using the PCMM for guidance, you determine that formalizing CI requires implementation in phases, and each phase has some association with a level in PCMM.

- **Phase 1: Document the Processes**

  To get started in PCMM, the first step is to document all processes. You decide the initial scope of effort will be limited to the actual manufacturing processes. In most cases on the production floor, you have found that the line supervisor has the best high-level understanding, so the line supervisors are your first stop.

  The line supervisors take you on a tour of their lines and describe the activity at each operation as the production operator does the work (a *Gemba walk* in Lean terms). You document the process in a flow chart, including the steps for each operation. You return to your desk and incorporate all these notes into an end-to-end flowchart file linking all operations together. After making a few trips back to the line to verify details, you find that most of the process is followed, but there are some *variations* at some operations, so you document these variations and the reason for the variations as well. You also find similar operations on multiple lines, but the steps followed in each line for these operations differ. This describes the conditions on production floors at PCMM Level 1.

- **Phase 2: Stabilize the Process**

  When looking at process interactions to determine the root cause, your training has shown that it is important to look for interactions that have significant correlations to the problem. However, you have found that if there is a lot of variation in the process, characterizing everything is very time consuming and confusing, and has the effect of diluting process interactions to the point where nothing seems to be significant. So, you decide that the next step in the CI initiative is to determine *a process* to follow and to stabilize that process (get everyone on the same line to follow the same process) with the goal of having a process that is repeatable and reproducible. You work with the supervisor of each line to determine a process that would work for their line, including decision points for variations that are needed and the criteria used to make decisions. You then train all the production operators on each line to follow "their process" and ensure that they understand this is not the "final" process; this is done only to measure the process reliably. In later stages, changes will be made to improve specific process parameters. After a few production runs, you find that the production operators are following the defined process for their line. This describes the conditions on production floors at PCMM Level 2.

- **Phase 3: Characterize and Improve Process Stability**

While you have been investigating issues in the process, you find that only when you have stabilized a process can the effort of measuring the process be beneficial. You determine that to engage in activity to characterize a process too early, the characterizations (determining the measurements) of the process change significantly from production run to production run. But, once a process is stable, it becomes easier to measure process capability (throughput, first-pass yield—FPY, cycle times, etc.) to determine what to improve. During this phase, you find that the majority of the effort is still to improve process stability; however, by removing variation in the process, you find that it also helps to improve process capability. When there is a stable general characterization, management decides which priority to improve (usually a choice between throughput and FPY), but you ensure that the management decision remains regarding this overall process (e.g., line throughput or line-level FPY). The mistake that many companies make is attempting to improve many things at the same time.

As stated in Chapter 3, when you have the general characterization, you then engage in a deeper analysis of process characteristics (detailed characterizing of a specific issue) looking for specific characteristics that closely correlate to the most significant issue determined by the Pareto chart.

In this phase, the production floor will go through iterations of improvement cycles (detailed in Chapter 3) and will revisit the process variations that were removed during process stabilization. This is necessary to ensure that localized knowledge (sometimes referred to as *tribal knowledge*) is not lost.

- **Phase 4: Change Focus to Process Improvement**

Recalling the iteration of the Deming cycle, the primary production process is defined and established as the "current practice." Using the Pareto chart, you determine the most significant issue to investigate. Depending on management decisions, the issue being investigated will start to change from removing process variation to improving general process characteristics (the throughput and FPY examples discussed earlier). This may require altering the "current practice" process to collect more detailed data and deeply evaluating the relationship of that data to the significant issue. This is done to achieve an understanding of which process characteristics have the greatest influence on the significant issue and to determine what range of those characteristics is best for the process goal. In following the practice of design of experiments, multiple iterations of alterations from current practice may be needed. Each time a specific alteration is implemented for a specific production run, the data is collected, and time is required to analyze the data. While the collected data is being analyzed, production continues using the current practice process (not the altered process). Once the preferred process characteristics are determined, a solution is defined to maintain them. This may be a simple change in equipment settings or a step in the process, or it may require a major upgrade in equipment. (This investigation may also take iterations of altering the current practice process in the same manner as before.) After a solution has been accepted, the current process is altered to use the accepted solution, which now becomes the new "current process."

When you have a better (more normalized) understanding of the characteristics of processes locally (within the line), you find similarities of processes across the boundaries of each line. As a result, you learn that specific people from each line are comparing notes and discussing problems as a group. During this phase, information from the analysis is shared as centralized knowledge bases are started and a CoE-like function starts to form. The more these "current practices" are shared across production boundaries, the more they become recognized and adopted as the company's "best practices." This describes the conditions at the plant level that are well established at PCMM Level 3.

- **Phase 5: Improve the Business and Support Processes**

  While solutions for process improvement are determined and implemented across company boundaries (development of the company's best practices), you realize that some of these solutions also require standardization and improvement in the business and support processes. This requires changes that extend into all four pillars of the MOM activity model. Best practices are defined for production planning, maintenance and calibration of equipment, operator training, and quality procedures. Any issues that use a mitigation strategy are implemented knowingly as the "best solution at the time." You establish management processes to continually review and manage these processes quantitatively through empirical investigation, and they are validated regularly as being "in control." You also ensure that mitigations are reviewed regularly to determine whether a more robust solution is available. You introduce using the MOM activity model as a template to help guide activities in this phase. When new processes are defined because of a new product or new production facilities, these best practices are implemented from the start, and new product introduction, or introduction of new equipment, becomes much more reliable. This describes the conditions at the plant level (also across multiple plants) at PCMM Level 4.

- **Phase 6: Strategic Planning of Major Improvements**

  In this phase of process improvement management, your company has achieved a very high level of process maturity and process throughput, and FPY (and any other key performance indicators—KPIs that are managed) is stable, reliable, and managed throughout the MOM scope of functionality. Senior management actively seeks input from your group regarding current production capacity and discusses high-level plans for changes to the manufacturing environment as a whole. At this phase of process maturity, you realize that most of the activity is related to major changes that are planned at the senior management level. This describes the conditions at the plant level at PCMM Level 5.

There is, of course, a close relationship between the phases of formalizing a CI program and PCMM because PCMM was developed for this purpose. However, it also must be recognized that business process management (BPM) and PCMM can (and should) be applied throughout the entire company and can be equally applied to any business or industry (not just manufacturing). Using manufacturing as a starting point simply makes good sense because process management has been a part of manufacturing since the days of Henry Ford and the original Ford manufacturing plant. It is important to ensure that the program does not end when you leave the manufacturing floor.

Methodologies in Continuous Improvement

In Chapter 3, I discussed using PCMM and introduced how it can be used to support process improvement methodologies such as the Theory of Constraints, Lean, Six Sigma, and ISO-9000. In this section, I discuss the differences between these methodologies and when they should be used; I also relate them to specific activities as presented in the phases listed earlier.

Many people have heard the saying, "If all you have is a hammer, everything looks like a nail," but what does it mean? The basic meaning is that if you have limited resources, you will always tend to take action that fits within the available resources. So, how does this apply to manufacturing? If the only CI methodology available to a company is ISO-9000, the company will try to solve all its problems using ISO-9000. If it has implemented Lean, it will try to apply Lean methods to everything. (Consultants are notorious for this mentality.) The issue is that, like a hammer, there are certain applications where Lean just does not fit.

In Chapter 7, I provided a simplified example of applying Hoshin Kanri in a manufacturing context. In that example, I presented activity from a "day-to-day improvement" context, which would fit within the Lean methodology (this is what Lean is for). I also presented an activity for "breakthrough improvement" that would fit within the Six Sigma methodology. The two methods are very complementary, which is why the industry is moving from certifications such as a Black Belt in one of these methods to a Black Belt in Lean-Six Sigma combining the two as opposed to using each methodology individually.

On the other hand, ISO-9000 stipulates the activities required to implement process standardization and helps to ensure that activity is applied to CI. It does not reflect on the methodology of analysis and improvement (for Lean and/or Six Sigma or developing an independent internal methodology), only that improvement is managed somehow. In this perspective, ISO-9000 fits into the methodology needed to step up in levels of PCMM (e.g., moving from Level 2 to Level 3).

Depending on the volume of product being manufactured, it can make sense to dedicate a specific line (or even a plant) to the production of that product. In this environment, Lean's "one-piece flow" was developed to make this model of a manufacturing line the most effective. If, on the other hand, production volumes are not that high, multiple products will have to share the same resources. In this environment, there will always be a single operation that will constrain the line's throughput. The Theory of Constraints is the management tool that works best in this scenario.

Understanding the use and limitations of management tools like ISO-9000, Lean, Six Sigma, and the Theory of Constraints is as important to manufacturing as understanding the use and limitations of a hammer or saw in construction. There is also value in understanding models like PCMM and ISA-95. Again, the correct use of each of these models as templates helps provide guidance to the CoE to measure the completeness of the process definition and the management of growth in process management. Using the same CoE for any program implementation of this kind helps ensure that each of these management tools is used effectively and that the knowledge from each program will build on that of the others.

Linking Process Data and System Data

In chapter 4, I discussed ensuring that the *process data* that contains the measurement of process performance, fallout, and other characteristics related to process health is linked correctly and incorporated into the system data of the IT systems that are monitoring production, but is this difficult to do?

When you look at data from a process perspective, things may seem fairly simple. The FPY is *a tally of all production units that have been through the process end-to-end without rework, without failing any tests or inspections, and by performing each operation in the process once.* This seems like a straightforward explanation of FPY. However, when you look at the history of production units that have completed the process from an IT perspective, you will find as an example, that some of the production units have been sampled by a quality engineer to determine a statistical characteristic of the process. Is this considered to be a normal part of the process? Were any of the production units at the low end of statistical acceptance (and maybe considered questionable), or were any retested and still passed for acceptance? For the process purists, if there is a deviation from the "normal process," it should not be considered as part of FPY. From a practical-minded perspective, if the production unit still passes all tests without intervention, it should be considered within FPY. These are all optional process considerations that must be evaluated. But how do you consider these concerns from a system data perspective?

In addition, relational system data records combine the data of multiple IT data objects to identify a specific process data object. In a relational database, these combined records can become very complex. If you are considering the scenario in the previous paragraph, there may be multiple configurations of system data that still have a common process data interpretation; however, that interpretation may require management decisions. These kinds of decisions must be based on a clear understanding of the processes, the configuration of master data within the system, and the possible impact of reporting and analysis using system data. This, again, stresses the importance of the CoE's role to accurately interpret reporting and analysis and to advise management on decisions.

When defining the reporting and analysis requirements, the following process must be followed.

1. There must be a clear understanding of the process outcome and a clear definition of what a "good product" looks like and what constitutes a complete product. As presented in one of the "Real-Life Experience" examples in Chapter 5, "the product" may only be the physical product that is being manufactured, or it may be the physical product with a report that identifies various performance characteristics and test results that describe the physical product's capability.

2. After the process has been defined, it would be of value to define how and where within the process information the product capabilities will be described. It is important to have a clear understanding of the context of the

process information to maintain the information's integrity (not to be confused with the system's data integrity).

3. Only after you have this understanding can you define how the IT system will capture and store that information. You must understand how the IT data elements will interact to define the process information properly.

4. When the IT data structure is clearly understood, you can then define the reporting and analysis structure while maintaining the information's integrity.

There are two primary causes of errors in reporting and analysis: manufacturing engineers not understanding the desired initial outcome (or being unable to define the outcome properly) and IT professionals taking a shortcut through the process and misinterpreting the initial purpose of the report or outcome of the process.

Connecting Dots

In the following sections, I provide expanded thoughts on topics I covered in earlier chapters.

A Final Reflection on MOM

In Chapter 5, I reviewed the ISA-95 explanation of MOM in detail. In that review, I explained the MOM activity model, referenced the parts of the ISA-95 Series of Standard that define the process data structures that must be a part of MOM, and described the four pillars of MOM. With all the components of the ISA-95 standards that describes MOM, I am surprised there is still a lot of confusion around what MOM is. Part of this confusion results from IT systems being advertised as "MOM systems" that have similar capabilities as MES.

So, what is the story here? Hopefully, I will clear this up in this section.

Figure 8-4, ISA-95 hierarchy of activities, illustrates the distinction between activities (Level 1 through Level 4), with Level 4 presenting the activities at the enterprise level for planning and logistics and Level 3 at the plant (or manufacturing floor) level for the execution and real-time management of the actual production process (order fulfillment).

This hierarchy of activities also indirectly defines the level of support that systems at these levels must provide and the interfacing that is required between these levels. Figure 8-5 depicts the relative amount of MOM support each system provides, as well as parts of the MOM activities supported by the interfacing layer (service-oriented application—SOA).

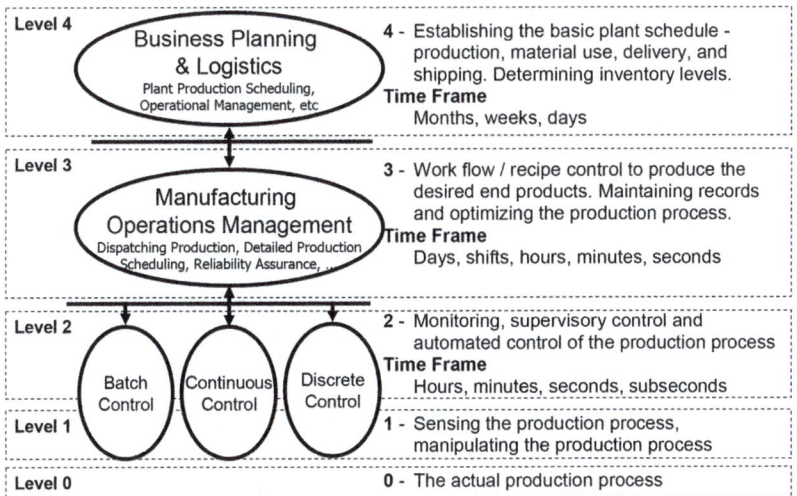

Figure 8-4. ISA-95 functional hierarchy of activities.

Source: Reproduced with permission from the International Society of Automation. ANSI/ISA-95.00.01-2010 (IEC 62264-1 Mod), *Enterprise-Control System Integration – Part 1: Models and Terminology*, figure 3.

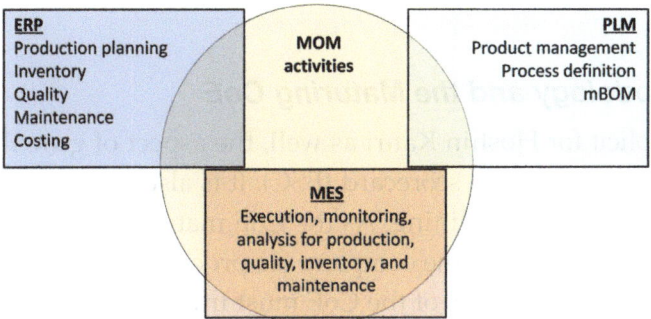

Figure 8-5. MOM activities provided by each system.

As part of developing IT systems that support these activity levels, most of this support has been developed within three main types of systems: enterprise resource planning (ERP), product lifecycle management (PLM), and manufacturing execution systems (MES).

In Chapter 6, I discussed the interfacing needed between these three systems using the SOA interfacing layer. I explained this layer further in Figures 6-6 and 6-7, in what I referred to earlier as *thin-layer* and *fat-layer* interfacing. As part of the interfacing requirements, *most* MES vendors have developed interfaces using SOA applications to ERP as a minimum, but with interfacing partitions that can be extended to PLM as well. The important point is that, in many cases, *the SOA layer already exists as part of the MES application suite.* In an attempt to differentiate a capability, some MES vendors have developed an SOA layer with the full support of interfacing for both ERP and PLM as well as the normalized data structure I referenced in discussing the fat-layer SOA (Chapter 6). In an effort to manage application updates and establish additional revenue streams by implementing an additional level of licensing, some companies have renamed the SOA layer as a *MOM system.* Because the functionality of interfacing to PLM can be developed for any implementation using the SOA applications already in use by many MES vendors and all the additional functionality of maintenance, scheduling, and quality have been added to either the SOA or the MES application itself, it is questionable whether renaming this application layer is required. An ERP system has a specific configuration (including the coding and database development) to support process manufacturing that is different from discrete manufacturing without changing the name of the system from "ERP." (There are also similar structural changes for all applications that support MOM.) Is it really necessary then, to rename an SOA application to being a "MOM application" just for the sake of marketing?

The manufacturing company's CoE must be aware of the real differences between these systems and provide proper guidance to the company's steering committee for system selection.

Strategic Methodology and the Maturing CoE

Although it is implicit for Hoshin Kanri as well, the aspect of growth and learning is an integral part of the balanced scorecard (BSC). It is also important is to ensure that the right people get the right training. As the CoE matures and gains greater knowledge of the internal workings of the company via process management and the implementation of initiatives, members of the CoE must investigate the newer technologies and "management tools" to understand if or when these new developments can be effective for the company. Although specific solution providers might introduce a new technological product, it is more important to understand the technology (or methodology, not necessarily the particular product) to understand the implications for manufacturing. The amount of effort directed to this understanding will also depend on the manufacturing company's strategic planning. (Will the company be an industry leader or a technology lagger but with better cost delivery?) The direction of the CoE's growth should be planned just as much as the manufacturing floor's growth.

An additional thought in ensuring the "right people" are trained is to ensure that the manufacturing and quality engineering teams also gain an understanding of the general structure of data that is available in the systems supporting manufacturing activities.

Real-Life Experience

One of the companies I consulted with on an MES implementation became highly protective of the system after the implementation went live. The implementation team, being almost exclusively IT, wanted to ensure that the configuration maintained its integrity. This resulted in a policy that only people on the IT team could access the MES configuration management function. None of the manufacturing engineering team members were given a user account in the system. This resulted in the manufacturing and quality engineers not gaining any knowledge of the system's capabilities or how to describe requirements for production support and reporting. For the first year of the implementation, manufacturing management was extremely disappointed with the system's capability to support production, and not knowing how the system worked, the manufacturing and quality engineers failed to provide requirements for reporting and analysis to support the company's CI program. Within the second year after the system's go-live, I was called back to review the implementation and explain why the system was not performing as expected. After reviewing the configuration and program management policies, I determined that the main problem was not the system but the inability of the manufacturing and quality engineers to properly explain reporting and analysis requirements that resulted from (as I explained it) the different languages the two teams were speaking. The engineering teams were expressing requirements in manufacturing terms, but the IT team was interpreting requirements in IT terms. I then presented the concept of an integrated CoE and recommended changes to drop the IT team's exclusive access to the system and provide training for the engineers to enable the team to develop a unified language and understanding. It was only after these recommendations were adopted that the company's CI program started making progress.

Information Integrity and Strategic Management Planning

The process for most strategic planning methods includes monitoring and documenting issues affecting the performance of manufacturing operations. Using the information that has been documented, senior management determines the direction and focus for operations to follow to support senior management's strategic direction. As with all analysis functions, the accuracy of the data and relevance of the information senior management uses will have an impact on the quality of the decisions made. Using an MES to monitor and report on the current state of manufacturing capability (as with using tactical scorecards in BSC) will help ensure the timeliness and accuracy of the data that is reported to senior management. In fact, most of the dashboards created to monitor strategic progress will likely come from the manufacturing floor and be reported from an MES. A mature CoE will help ensure that the information interpretation is both correct and relevant. This, in turn, leads to better prioritizing of initiatives and better estimation of costs and timing for projects and significantly improves the capability to create and maintain a budget.

More on Tools of the Trade

In Chapter 1, I discussed the monthly operational review (MOR) as one of the tools of the trade for manufacturing management. Also, I presented the scenario of operations staff working for up to a week to summarize the current state of their manufacturing lines and the issues they are facing. In addition, I discussed using standardization as a means to minimize both the complexity and frequency of management decision-making. In Chapters 2 and 4, I discussed using the exceptional amount of data that has become available via Industry 4.0 and the technology that is becoming available with devices from IIoT.

It then surprises and concerns me how few companies will take advantage of the real-time reporting capability of an MES and have all their MOR reports made available via their MES. All the data that is needed for the MOR comes from an MES, and the system does not filter out the issues that staff will frequently try to hide (and maybe that is the point). If the company is looking to gain confidence in its system (processes, people, and IT stack), it will have to get used to the transparency that goes with real-time reporting. These "unexpected" reporting concerns are simply another indication of variation that must be resolved.

Thoughts on Inventory and Process Management

Inventory management is tied to the company's warehouse locations, inventory that is externally in transit, and inventory that is purchased but may not have been shipped yet, and ERP is usually *aware* of how much inventory of each material is available. The ERP system also is aware of where each segment of inventory is located (and if it is in multiple locations), as well as how long the inventory has been available at each location. In many cases, the ERP system may also be aware of the batch IDs that are available at each location. In many manufacturing companies, inventory management is a major part of the activity of operations. These activities include cycle counting (to ensure accurate quantities), movement of material from secondary storage locations to primary locations used to create Kanbans, or other types of kits that must be prepared for movement to the production floor. Finally, the ERP system may include functionality that helps create pick lists to ensure efficient use of the movement of people and equipment.

One issue that is surprisingly common in manufacturing plants is that material has been moved out onto the production floor (delivered to a workstation for use), but the ERP system still shows this material as "available." This is despite the fact that after material is moved to a workstation, the expectation is that the defined quantity that was moved is usually viewed from a practical sense as being "allocated" to the product that the production line is producing. In some cases, the entire production floor is designated as a single inventory location. (After material is moved to the floor, it is unknown from a system perspective where on the production floor the material was moved to.) This situation frequently results from "relaxed inventory practices" and leads to a lack of system visibility of the inventory (which in turn leads to excess inventory being moved to the floor or being purchased). These companies also rely heavily on expediters to keep track of where inventory is on the floor and to distribute it where necessary. What is not recognized is that this scenario is prone to the same kind of human errors found in manufacturing that manufacturing is trying to correct with CI initiatives.

Most MES provide some support to inventory management on the production floor with functionality that supports line-side stocking (LSS). Depending on the details determined by the configuration, LSS can track material quantities, batch IDs of each material allocation, and inventory consumption during the manufacturing process. But implementing LSS in an MES implementation will not be a fix for bad inventory management. There is an adage that "to err is human, but to really mess things up, you need a computer." Before implementing any aspect of automation, it is first necessary to understand and standardize the process that is being automated. If a process has not been standardized before being automated, it will simply speed up the human error that is causing problems. In this case, is it the computer that is "causing the mess" or the people who implemented automation before the process was standardized?

> "The whole is greater than the sum of its parts."
>
> —Aristotle

My Final Words

Throughout this book, I have discussed the advantages to a manufacturing company of developing and managing a CoE for MOM. Many books have been written about using teams, and they include quotes such as "the whole is greater than the sum of its parts." What is not always explained in the discussion on teams is that diversity of knowledge is just as important in *creating the whole* as the contribution of multiple inputs. With the implementation of many IT systems for the manufacturing floor, it is not always recognized that at the manufacturing operations level (Level 3 in the ISA-95 standards), the primary purpose of the system is to create a virtualization of the manufacturing floor (referred to as a *digital twin*) and provide feedback (and insight) into the "current state" of that manufacturing floor at any time. In addition, these systems must provide notification of events that happen on the manufacturing floor as they are happening. The closer the virtualization is to replicating the manufacturing floor, the better the information the system can provide.

It takes a lot of effort and diverse knowledge to configure and maintain the virtualization presented by the systems, depending on the complexity of the processes to manufacture the products as well as the processes to support manufacturing. In this book, I have indicated the knowledge required by and the structure of the team that will define these processes, and I have provided some insight into the activities this team will participate in. This insight includes knowledge from each of the four aspects of MOM (the four pillars), from each of the five levels in the ISA-95 standards (including Level 0), and from the company's IT and data analytics teams. Regardless of what the team is named (I refer to it as a CoE), the diversity of that knowledge and the opportunity to work as a single team will be essential and should never be underestimated.

Another concept I have discussed (sometimes at length) is using "the right tool for the job" and ensuring that these "tools" are supported by the knowledge of how to use them. In the scope of manufacturing, using tools such as PCMM, ISA-95, Lean, and ISO-9000 is important to the effectiveness of that team. However, the "right tools" will also likely change within the next few years. Ensuring that manufacturing companies can keep up with these changes is another reason for supporting the CoE concept. I hope that I have shown the benefits in the manner that these "tools" can interact and support the overall effectiveness of the manufacturing floor.

> "What the caterpillar calls the end, the rest of the world calls a butterfly."
>
> —Lao Tzu

Throughout the book, I have also discussed the importance of change management. Change management is not random change but methodically investigating and adopting change. In addition, a company's culture must recognize that change is neither good nor bad. A company must change over time to keep up with consumer demands and changes in industry. Embracing and managing that change is what is important. As stated in a few chapters, variance (random change) is bad for manufacturing management. To effectively investigate and adopt change, variance in processes should be minimized; however, minimizing change should not be reflected as an aversion to proper change management.

I hope that the readers of this book will find my explanations of the concepts covered to be insightful and helpful in their investigation of MOM (both generally and in accordance with the ISA-95 standards) and their investigation of MES. I welcome feedback and can be contacted via LinkedIn at https://ca.linkedin.com/in/vokeygcm.

Abbreviations

AI	artificial intelligence
ASN	advanced ship notice
ATO	assemble to order
BOM	bill of material
BPM	business process management
CI	continuous improvement
BSC	balanced scorecard
CMM	capability maturity model
CoE	Center of Excellence
COO	chief operating officer
COTS	commercial-off-the-shelf
DHR	device history report
DoE	design of experiment
ERP	enterprise resource planning
FPY	first-pass yield
HMI	human-machine interface
IIoT	Industrial Internet of Things
ISA	International Society of Automation
IT	information technology
KPI	key performance indicator
LCL	lower control limit
LIMS	laboratory information management system
LSS	line-side stocking
mBOM	manufacturing bill of material

MES   manufacturing execution system
MESA   Manufacturing Enterprise Solutions Association
MOM   manufacturing operations management
MPS   master production schedule
MRP   materials requirements planning
OEE   overall equipment effectiveness
OEM   original equipment manufacturer
OLE   Object Linking and Embedding
OPC   OPC server (Open Platform Communication or OLE for Process Control)
OT   operational technology
PCMM   process capability maturity model
PLM   product lifecycle management
S&OP   sales and operations planning
SaaS   software-as-a-service
SCADA   supervisory control and data acquisition
SCM   supply chain management
SDLM   software development lifecycle management
SI   system integrator
SOA   service-oriented application
SOP   standard operating procedure
SPC   statistical process control
TOC   Theory of Constraints
TPM   total productive maintenance
UCL   upper control limit
UML   Unified Modeling Language
WIP   work in process

Glossary

BOM: bill of material. A list of materials and the quantity of each that are used to make one unit of a product.

BOP: bill of process. A list of operations and/or tasks that must be performed to make a production unit.

blue printing. The part of an implementation project that analyzes the current state of a manufacturing plant's operation and determines the vision of a future state. It then compares that current state to the vision of the future state and creates a roadmap of changes to achieve the future state.

DFM: design for manufacturing. A design methodology that focuses on designing a product in a manner that optimizes the capability to manufacture the product.

ERP: enterprise resource planning. The aggregate of processes, systems, and data used to manage the ongoing capabilities of a company as a whole (or enterprise).

IIoT: Industrial Internet of Things. The interconnection of sensors, instruments, and other devices networked together with industrial applications running on various computing devices to support manufacturing, energy management, and other industrial purposes.

IoT: Internet of Things. An extension of Internet connectivity into physical devices and everyday objects (cyber-physical devises). These devices are embedded with electronics, Internet connectivity, and other forms of hardware (such as sensors); they can communicate and interact with other devices over the Internet and can possibly be remotely monitored and/or controlled.

ISA-95: International Society of Automation ISA-95 Series of Standards. These standards contain an information technology (IT) model linking the processes of the shop floor to the top floor of manufacturing.

just-in-time. A process in which manufacturing scheduling is tightly managed to minimize any gaps in time from the launch of production and delivery of any unit to the expected demand of the end product.

LSS: line-side stocking. An inventory stocking methodology that places small quantities of material stock at a workstation-level inventory location. It is used to ensure easy access to material that will be needed within a short time window.

M2M: machine-to-machine. Early concepts of IIoT used to reference the communication events directly between two or more machines without a central database system.

mBOM: manufacturing bill of material. A list of the raw material (components) required to manufacture a single unit of the product being manufactured, including data that is important to the manufacturing process, such as operation of consumption, process, or genealogy data to be collected.

MES: manufacturing execution system. A system used to monitor and control the activities and resources of the manufacturing floor in a company (or enterprise).

MESA: Manufacturing Enterprise Solutions Association. An international not-for-profit community dedicated to improving business and manufacturing practices through optimized application and implementation of information technology (IT) and best management practices.

MOM: manufacturing operations management. The aggregate of processes, systems, and data used to manage the ongoing activities on the manufacturing floor as defined by the ISA-95 MOM activity model.

NPI: new product introduction. The aggregate of processes, systems, and data used to manage the ongoing activities of bringing a product from concept to market.

OT: operational technology. From a manufacturing perspective, the hardware and software that directly acts on a production unit as part of the manufacturing transformation process.

OEM: original equipment manufacturer. A manufacturing organization that makes a consumer end product from parts and subassemblies ordered from other manufacturing organizations.

plant. Manufacturing facility. A reference to a building (or multiple buildings) in which manufacturing takes place.

PLM: product lifecycle management. The aggregate of processes, systems, and data that are used to manage the ongoing lifecycle of a product from concept and release through to product obsolescence and discontinuation.

readiness. The status of equipment, material, or process that references its availability, operability, and suitability for use. This can include scheduled availability, having an operator available to use it, and the "fit for use" for the task including calibration/maintenance and the capability to perform as required.

SCM: supply chain management. The aggregate of processes, systems, and data that are used to manage the ongoing activities within an enterprise for the receipt of raw materials from vendors to the delivery of end product to customers. (Note: In some situations, this may include the extended supply chain over multiple companies.)

Theory of Constraints. A method of system and process analysis that states that all processes will have a single constraining function and prescribes methods to manage and improve a process based on the defined constraint.

work-in-process. One or more units of a product in the process of being assembled or fabricated.

Index

Printed and bound by CPI Group (UK) Ltd, Croydon, CR0 4YY

13/05/2026

14879213-0002